PLATE I
(*Frontispiece*)

(A7107—7108)

SADDLE FOLD IN HYTHE BEDS, DRYHILL, SUNDRIDGE

NATURAL ENVIRONMENT RESEARCH COUNCIL

INSTITUTE OF GEOLOGICAL SCIENCES

MEMOIRS OF THE GEOLOGICAL SURVEY OF GREAT BRITAIN
ENGLAND AND WALES

Geology of the Country around Sevenoaks and Tonbridge

(Explanation of One-inch Geological Sheet 287, New Series)

by

H. G. DINES, A.R.S.M., A.M.INST.C.E.,
S. BUCHAN, PH.D., S. C. A. HOLMES, M.A.,
and C. R. BRISTOW, PH.D.

with contributions by

Sir James Stubblefield, D.Sc., F.R.S., R. V. Melville, M.Sc.,
F. W. Anderson, D.Sc., F.R.S.E., R. Casey, Ph.D., D.Sc., H. C. Ivimey-Cook, Ph.D.,
M. Mitchell, M.A., C. J. Wood, B.Sc., A. C. Benfield, B.Sc.,
K. C. Dunham, D.Sc., S.D., F.R.S., M.I.M.M.,
P. A. Sabine, Ph.D., F.R.S.E., M.I.M.M.,
F. H. Edmunds, M.A. and J. N. Carreck

LONDON
HER MAJESTY'S STATIONERY OFFICE
1969

The Institute of Geological Sciences
was formed by the
incorporation of the Geological Survey of Great Britain
and the Museum of Practical Geology
with Overseas Geological Surveys
and is a constituent body of the
Natural Environment Research Council

SBN 11 880067 1

Printed in England for Her Majesty's Stationery Office
by Hull Printers Ltd., Willerby, Hull, Yorks.

PREFACE

THE AREA of Sevenoaks (287) Sheet of the New Series One-inch Geological Survey map forms part of that of the Old Series One-inch Geological Survey map Sheet 6 (Bromley). The first edition of this map, for which W. Whitaker surveyed the Chalk and F. Drew the beds below the Chalk, was published in 1864, but it showed only the Solid formations. Later the Drift deposits north of the Chalk escarpment were surveyed by W. Whitaker, T. McK. Hughes, F. J. Bennett, W. A. E. Ussher and C. E. Hawkins and those of the remainder of the area by C. Le Neve Foster, W. Topley, W. B. Dawkins and C. E. Hawkins, thus permitting the publication, in 1886, of a Drift edition. Some revised boundary lines were added to a third edition in 1889.

The resurvey of the main part of Sheet 287, on the scale of six-inches to a mile, carried out by Dr. S. Buchan, the late H. G. Dines, Mr. S. C. A. Holmes and Dr. F. B. A. Welch, started in 1930 under the direction of the late H. Dewey as District Geologist and was completed in 1936 under the late C. E. N. Bromehead. Small areas had already been surveyed by Messrs. Dewey and Bromehead in 1913–14 and in 1921 during the resurvey of Dartford (271) Sheet to the north. Minor additions in the south-eastern corner were made by the late F. H. Edmunds in 1946. The Hastings Beds were revised by Dr. C. R. Bristow in 1965 following the completion of his work on the Hastings Beds of the adjoining Tunbridge Wells (303) Sheet to the south.

Literature on the geology of the district is considerable. Many references may be found in the Geological Survey Memoirs 'The Geology of the London Basin,' by W. Whitaker, 1872, 'The Geology of the Weald' by W. Topley, 1875, and 'The Cretaceous Rocks of Britain,' Vols. I-III, by A. J. Jukes-Browne, 1900–4, while most of the older records of wells in the area are given in 'The Water Supply of Kent', by W. Whitaker, 1908, and 'The Water Supply of Surrey' by W. Whitaker, 1912. The information from wells has been brought up to date and published in the Well Catalogue Series of the Geological Survey (Reigate (286) and Sevenoaks (287) sheets, Cooling 1968). Other references appear in various important papers in the *Quarterly Journal of the Geological Society* and the *Proceedings of the Geologists' Association*.

Publication of Sheet 287, in 1950, was delayed by World War II and its after-effects; indeed the map was approaching the publication stage, when the preparatory work was destroyed by enemy action. Other unavoidable factors have further delayed the issue of the Memoir, and availability of officers has been reduced by retirement and transfer to other duties. Advantage of the delay has been taken to include data obtained since the completion of original field mapping, and to bring this account into line with a revised One-inch sheet in preparation. Sir James Stubblefield and Mr. M. Mitchell have contributed notes on the palaeontology of the concealed Palaeozoic strata of the Penshurst trial borehole for oil, and Dr. F. W. Anderson, Dr. R. Casey, Dr. H. C. Ivimey-Cook, and Mr. R. V. Melville on the Mesozoic; myself (when Chief Petrographer) on the petrography of the Wealden Beds and Lower Greensand; Dr. P. A. Sabine on the petrography of the Carboniferous Limestone and of the pebbles of the Limpsfield Gravel; Dr. Casey on the palaeontology of the Gault; Mr. C. J. Wood on the palaeontology of the Chalk; and Mr. A. C. Benfield on the Water Supply, incorporating earlier observations of Dr. S. Buchan. Dr. Casey has identified many of the fossils named, except those quoted from other authors. We are grateful to Dr. J. H. Callomon and Dr. J. C. W. Cope for determinations of ammonites from the Penshurst boreholes. Mr. J. N. Carreck, of Queen Mary College, London, has kindly provided notes on the Pleistocene fissures at Ightham, including a list of the fauna of which the Mollusca have been checked by Dr. M. P. Kerney, of the Imperial College of Science and Technology, London. The Memoir was edited by the late F. H.

Edmunds, and by Mr. S. C. A. Holmes and Dr. C. R. Bristow who have also made additional contributions. Official photographs were taken by Messrs. J. Rhodes and J. M. Pulsford.

Thanks are given to many land and quarry owners for assistance in examining sections, to the Eastern Gulf Oil Company Ltd. for permission to include data obtained from their trial borehole at Penshurst and to the geological staff of the Ministry of Housing and Local Government for data on working clay-pits, stone quarries, etc., included in Chapter VIII of this Memoir; the late Sir Edward Harrison rendered valuable help both in the field and in connexion with the work of his father, Benjamin Harrison.

The late G. A. Peet kindly provided the surveyed plan of underground quarries near Westerham reproduced in Figure 6.

K. C. DUNHAM
Director
Institute of Geological Sciences
Exhibition Road
South Kensington
London S.W.7
26th June, 1969

CONTENTS

ILLUSTRATIONS

TEXT FIGURES

EXPLANATION OF PLATES

PLATE   I   Saddle fold in Hythe Beds, Dryhill, Sundridge.

A number of sharp anticlinal folds, or 'saddles', usually with approximately east–west axes, affect the Lower Greensand of the Westerham–Sevenoaks area. These appear to be relatively superficial adjustments to pressure from the south. This pit is now disused but the saddle fold can still be seen although talus partially obscures the face. (A 7107–7108)[1]   *Frontispiece*

FACING PAGE

PLATE   II A. Chalk escarpment and Gault vale, Wrotham.

The Upper Chalk caps the escarpment, from the crest of which the Middle Chalk extends to the lower of the two parallel hedge lines. Lower Chalk and Gault occupy the undulating ground in the middle distance, while Folkestone Beds, exposed in a sandpit, are in the foreground. The alluvium of the Darent crosses the view below the line of poplars. (A 5352)

B. View towards Penshurst from below the Lower Greensand escarpment.

The undulating tract of country in the foreground (on which can be discerned the route for the Sevenoaks By-pass) is occupied by the Weald Clay. The high ground (Ashdown Forest, etc.) in the middle distance is formed by the Hastings Beds and the Chalk escarpment of the South Downs is visible in the far distance on the right. (A 10271)   4

PLATE   III A. Basal beds of the Ardingly Sandstone, near Speldhurst.

Except at its base, the Ardingly Sandstone has lost its typical massive facies and appears as thin flaggy beds. This is a common occurrence in weathered man-made exposures. (A 6812)

B. Ardingly Sandstone near Chiddingstone Hoath.

Massive Ardingly Sandstone overlying thinner bedded sandstone; road cutting due north of Hoath House (formerly Batts Farm) 1 mile east of Markbeech. The vertical joints, cross bedding and honeycomb weathering are typical of this horizon at the top of the Lower Tunbridge Wells Sand. (A 10273)   32

PLATE   IV A. Joints in Ardingly Sandstone; Moorden Farm, near Penshurst Station.

The Ardingly Sandstone is usually regularly jointed by two systems at right angles to each other and perpendicular to the bedding planes. The sandstone is accordingly separated into large more or less cuboidal blocks. Honeycomb weathering, also characteristic of this horizon, is clearly seen in this disused quarry. (A 10274)

---

[1]Reference number of photograph in the Geological Survey collection of geological photographs.

LIST OF SIX-INCH MAPS

The following is a list of six-inch geological maps included in the One-inch geological map, Sheet 287, with the initials of the surveying officer and date of the survey.

The names of the officers are as follows: C. N. Bromehead, S. Buchan, H. Dewey, H. G. Dines, S. C. A. Holmes, F. B. A. Welch and C. R. Bristow.

Manuscript copies of the maps are deposited for public reference in the Library of the Institute of Geological Sciences. They contain more detail, largely in the form of annotations, than appears on the one-inch map.

SURREY

21 N.W. (part of)	Downe	..	.. H.D.	1913 and S.C.A.H. 1936–8
21 S.W. (part of)	Tatsfield	..	.. S.B.	1931 and S.C.A.H. 1936
28 N.W. (part of)	Titsey	..	.. S.B.	1931 and S.C.A.H. 1936
28 S.W. (part of)	Limpsfield	..	.. S.B.	1931
36 N.W. (part of)	Marlpit Hill	..	.. S.B.	1932
36 S.W. (part of)	Haxted	..	.. S.B.	1932–3
43 N.W. (part of)	Dormans Land		.. S.B.	1935

KENT

28 N.W. (part of)	Downe	..	.. H.D.	1913–4 and S.C.A.H. 1936–8
28 S.W.	Tatsfield	..	.. S.B.	1931 and S.C.A.H. 1936–8
28 N.E. (part of)	Halstead	..	.. H.D.	1914 and S.C.A.H. 1936
28 S.E.	Knockholt	..	.. S.C.A.H.	1936
29 N.W. (part of)	Shoreham	..	.. H.D.	1914 and S.C.A.H. 1936
29 S.W.	Otford	..	.. H.G.D.	1932 and S.C.A.H. 1936–7
29 N.E. (part of)	Kingsdown	..	.. C.N.B.	1921
29 S.E.	Kemsing	..	.. S.C.A.H.	1936
30 N.W. (part of)	Stansted	..	.. C.N.B.	1921 and S.C.A.H. 1936
30 S.W.	Wrotham	..	.. S.C.A.H.	1936
30 N.E. (part of)	Birling	..	.. C.N.B.	1921 and S.C.A.H. 1936–7
30 S.E. (part of)	Offham	..	.. S.C.A.H.	1936
39 N.W.	Westerham	..	.. S.B.	1931 and S.C.A.H. 1936
39 S.W.	Crockham Hill		.. S.B.	1931–6
39 N.E.	Brasted	..	.. H.G.D.	1932
39 S.E.	Ide Hill	..	.. H.G.D.	1932
40 N.W.	Sevenoaks	..	.. H.G.D.	1932
40 S.W.	Sevenoaks Weald	..	.. H.G.D.	1932
40 N.E.	Ightham	..	.. H.G.D.	1936
40 S.E.	Shipbourne	..	.. S.B. & H.G.D.	1936
41 N.W.	Claygate Cross	..	.. H.G.D.	1936
41 S.W.	West Peckham	..	.. F.B.A.W.	1931 S.B. and H.G.D. 1936
41 N.E. (part of)	St. Leonard's Street	..	.. H.G.D.	1936
41 S.E. (part of)	Nettlestead	..	.. F.B.A.W.	1931 and H.G.D. 1936
49 N.W.	Marlpit Hill	..	.. S.B.	1932–3
49 S.W.	Edenbridge	..	.. S.B.	1932–5
49 N.E.	Four Elms	..	.. S.B.	1935–6
49 S.E.	Chiddingstone	..	.. S.B.	1935–6
50 N.W.	Hildenborough Station		.. S.B.	1936
50 S.W.	Leigh ..	..	.. S.B.	1935
50 N.E.	Hildenborough	..	.. S.B.	1936
50 S.E.	Tonbridge	..	.. S.B.	1936
51 N.W.	Hadlow	..	.. F.B.A.W.	1931
51 S.W.	Capel ..	..	.. F.B.A.W.	1931
51 N.E. (part of)	East Peckham		.. F.B.A.W.	1931

KENT

51 S.E.	(part of)	Paddock Wood	..	F.B.A.W.	1931
59 N.W.	(part of)	Dry Hill Camp	..	S.B.	1933–5
59 N.E.	(part of)	Cowden	.. ..	S.B.	1935
60 N.W.	(part of)	Speldhurst	.. ..	S.B.	1935
60 N.E.	(part of)	Southborough	..	S.B.	1936
61 N.W.	(part of)	Lower Green	..	F.B.A.W.	1931
61 N.E.	(part of)	Brenchley	.. ..	F.B.A.W.	1931

The revision survey of the Hastings Beds was carried out on the National Grid six-inch maps.

TQ 34 S.E.	(part of)	Lingfield	..	..	..	C.R.B.	1965
TQ 44 S.W.		Dormans Land	..	..	C.R.B.	1963–5
TQ 44 S.E.		Cowden	..	..	..	C.R.B.	1963–5
TQ 44 N.E.	(part of)	Hever	..	..	..	C.R.B.	1965
TQ 54 S.W.		Penshurst	..	..	..	C.R.B.	1963–5
TQ 54 N.W.	(part of)	Chiddingstone Causeway	..	C.R.B.	1965
TQ 54 N.E.	(part of)	Tonbridge	..	..	C.R.B.	1965
TQ 54 S.E.		Southborough	..	..	C.R.B.	1964–5
TQ 64 S.W.		Pembury	..	..	C.R.B.	1964–5
TQ 64 N.W.	(part of)	Tudeley	..	..	C.R.B.	1965
TQ 64 N.E.	(part of)	Paddock Wood	..	..	C.R.B.	1965
TQ 64 S.E.	(part of)	Brenchley	..	..	..	C.R.B.	1965

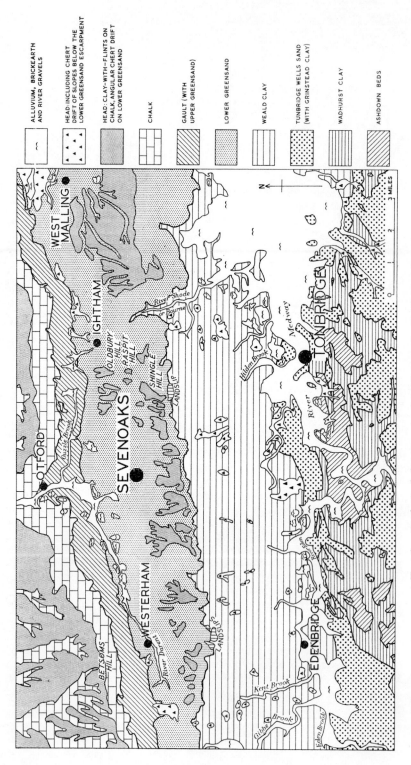

ALLUVIUM, BRICKEARTH AND RIVER GRAVELS

HEAD:INCLUDING CHERT DRIFT OF SLOPES BELOW THE LOWER GREENSAND ESCARPMENT

HEAD:CLAY-WITH-FLINTS ON CHALK, ANGULAR CHERT DRIFT ON LOWER GREENSAND

CHALK

GAULT (WITH UPPER GREENSAND)

LOWER GREENSAND

WEALD CLAY

TUNBRIDGE WELLS SAND (WITH GRINSTEAD CLAY)

WADHURST CLAY

ASHDOWN BEDS

Fig. 1. *Geological sketch-map of the Sevenoaks–Tonbridge district*

Chapter I

INTRODUCTION AND TABLE OF STRATA

THE DISTRICT described in this Memoir lies within the county of Kent, except for a narrow strip of Surrey along part of the western margin.

The solid rocks form part of the northern limb of the Wealden dome; outcrops of the various formations trend parallel with the major axis of the dome and cross the area in an east–west direction, the oldest formation exposed, the Hastings Beds, being in the south, and the newest, the Chalk, in the north. The general distribution of the formations occupying the surface is shown in the geological sketch-map, Fig. 1; a list of formations is given in the Table on pp. 6–7.

The outstanding physical features (illustrated in Fig. 2) are the prominent escarpments of the Chalk and the Lower Greensand. The Chalk escarpment, a view of which is shown in Plate IIA, forms part of the North Downs; it maintains a fairly constant level along its crest of between 700 and 750 ft above Ordnance Datum, but rises to 800 ft at Betsom's Hill and to 878 ft at Botley Hill on the western edge of the district; it is breached by the gap of the River Darent at Otford which has been cut down to about 200 ft O.D.

The Lower Greensand escarpment is of rather lower elevation than the Chalk escarpment, and is a more uniform feature across this area than its continuations to the east and west, possibly owing to the persistent development of hard ragstone and especially to the chert layers which form the upper horizon of the Hythe Beds and occupy the escarpment crest practically all the way across the district.

In the west, the escarpment is low, and is breached by the valley of a small stream flowing southward through Limpsfield and by a wind-gap at a level of 480 ft O.D. south of Limpsfield Common; eastward, it rises above 600 ft O.D. and continues approximately at that level as far as Wilmot Hill, with one peak above 800 ft O.D. at Toy's Hill. The small southward-flowing River Bourne, locally known as the Shode (Bennett 1907, p. 2) occupies a large gap, cut down to 150 ft O.D. at Plaxtol, beyond which the scarp rises to 550 ft O.D. in Mereworth Woods and then falls gradually eastwards. This well-defined feature overlooks the low ground of the Wealden Beds to the south, and landslips of Lower Greensand over the underlying clays are of relatively frequent occurrence. In places a second escarpment has been formed, owing to the local presence of chert layers at the top of the Lower Greensand. This commences near Seal and extends eastward to include Ightham Common and Oldbury Hill. The highest point of Ightham Common is locally known as Raspit Hill,

1

a name not appearing on the Ordnance Survey maps. South of Mereworth a prominent, dome-shaped hill rising above 350 ft O.D. is composed of an out-lying mass of partially disturbed Hythe Beds lying south of the main escarpment.

Between the Chalk and Lower Greensand escarpments lies the Vale of Holms-dale, about four miles wide, its northern slopes being the scarp face of the North Downs and its southern the dip-slope of the Lower Greensand, Gault lying along the floor.

South of the Lower Greensand outcrop the country is generally low lying, and is roughly bisected by the eastward-flowing rivers Eden and Medway, which follow approximately the junction between the Weald Clay and the Hastings Beds (Figs. 1 and 2). The Weald Clay country slopes gently south-wards from the Lower Greensand escarpment towards the rivers, while, on the south, country of rather higher relief and steeper slopes, formed by the Hastings Beds, rises gradually southward towards the High Weald.

The map area is mainly agricultural land and woodland, although extensive residential districts have grown around the towns and villages, notably at Tonbridge, Paddock Wood, Sevenoaks, Westerham, Riverhead, Dunton Green, Otford, Kemsing, Borough Green and Hildenborough.

The distribution of woods and commons and of agricultural land is governed to a considerable extent by geological factors. Much arable land is present on the Chalk, interspersed with patches of woodland: often hanging woods have been planted on the steep sides of coombes. Pasture occurs mainly on clay lands, especially on the Weald Clay, but this formation also supports arable land particularly where spreads of drift are present. The Hastings Beds country affords an example of land once heavily wooded that has now been largely cleared for agriculture; a mosaic of small patches of woodland, often in angular, geometric shapes, is suggestive of remnants after the making of rectangular clearings.

The Lower Greensand supports little agricultural land except locally, as in the valley of the River Bourne, south of Ightham; its outcrop is covered largely by woods and heathy commons and in recent years by residential building estates. Commons include Seal Chart with Ightham Common and Oldbury Hill; Hubbard's Hill, Bayley's Hill and Goathurst Common, south-west of Sevenoaks; the three Charts and Crockhamhill Common, south of Westerham; and Limpsfield Common.

The greater part of the map area drains into the Medway catchment with that of its tributary the River Eden. These rivers drain the whole of the area south of the Lower Greensand escarpment, as well as the eastern and western parts of the Vale of Holmsdale. The central part of the Vale of Holmsdale, however, constitutes the head of the River Darent catchment, which includes the downs above Kemsing. This river flows northward into the Thames Valley through the Darent gap at Otford. The Chalk downs of the north-west corner of the district are traversed by a number of dry coombes within catchments of the rivers Wandle, Ravensbourne and Cray. The distribution of the water-sheds is shown in Fig. 2.

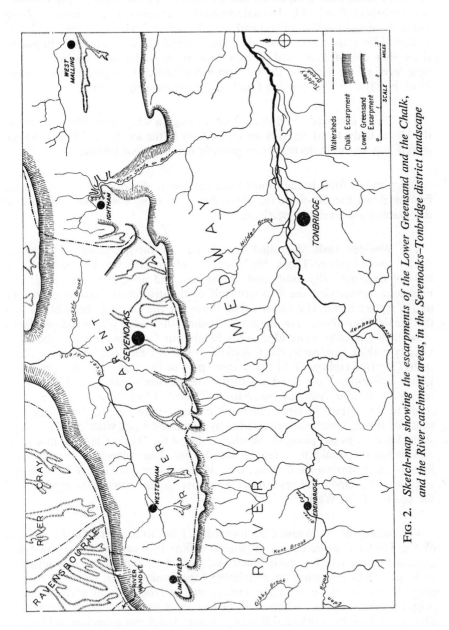

Fig. 2. *Sketch-map showing the escarpments of the Lower Greensand and the Chalk, and the River catchment areas, in the Sevenoaks–Tonbridge district landscape*

B

Although the Darent probably originated as a normal consequent stream, the headwaters of which, within the Weald, were captured by the Medway, the encroachment of the Medway system into Holmsdale, both on the west (Gibb's Brook, near Limpsfield) and on the east (River Shode, near Ightham) was probably influenced by Head deposits that for a period may have completely blocked parts of the Vale of Holmsdale and interfered with its normal drainage. Trains of later Head on the Weald Clay, particularly southward of Limpsfield, show that the Medway had already lowered its valley well towards present base-level at the time of their formation (Dines, Hollingworth and others 1940, p. 221). Further erosion finally dissected the original fans.

The Shode eastward of Ightham flows through a steep gorge where it crosses the Hythe Beds. Head deposits on Weald Clay to the south are similarly related to this gap and were derived from gravelly deposits near Ightham.

Sharp saddle folds in the Lower Greensand have diverted obsequent streams flowing down the dip-slope for short distances along the strike and affected the Darent itself near Westerham.

The Wealden Beds, deposited in a lagoon of varying salinity (Anderson, Bazley and Shephard-Thorn 1967), consist of an alternating series of thick beds of fine, buff silty sands and sandstones, and grey, laminated clays and mudstones, occupying the southern half of the district and covering an area of 119 square miles. The major subdivisions are the Hastings Beds below and the Weald Clay above. The former is further subdivided into Ashdown Beds, Wadhurst Clay and Tunbridge Wells Sand with Grinstead Clay in the west. Locally a thick sandstone, the Cuckfield Stone, is developed within the Grinstead Clay (see table p. 7). In the western part of the map area the uppermost Lower Tunbridge Wells Sand is a distinctive massive sandrock and is mapped separately as the Ardingly Sandstone (Gallois 1966). The Ashdown Beds come to the surface in large inliers, south and south-west of Tonbridge and south of Hever, which are surrounded by Wadhurst Clay, while the Tunbridge Wells Sand crops out along the southern edge of the Weald Clay and in several outliers on the southern margin of the district. The outcrop of the Hastings Beds is much broken by faulting, and the consequent patchwork of outcrops of clays, silty sands and sandstones gives rise to the varied characteristics of the centre of the Weald.

The outcrop of the Weald Clay formation, lying to the north of the outcrop of the Hastings Beds, covers 69 square miles; it generally forms low-lying ground, but the thin lenticular limestones, sandstones and ironstones give rise to low features imparting an undulating topography to the Weald Clay outcrop. In the central part of the district the outcrop rises to a considerable height up the scarp-face which is capped by the Lower Greensand, where it is deeply furrowed by southward-flowing streams, which rise as springs at the base of the sands.

The Lower Greensand includes the Atherfield Clay, the Hythe Beds, the Sandgate Beds and the Folkestone Beds. It occupies a strip of country gradually

(A 5352)

A. CHALK ESCARPMENT AND GAULT VALE, WROTHAM

PLATE II

B. VIEW TOWARDS PENSHURST FROM BELOW THE LOWER GREENSAND ESCARPMENT

(A 10271)

widening from just over a mile on the west to nearly six miles on the east. The outcrops of the four divisions extend continuously across the district and the total area covered is 54 square miles. Much of the high ground of the Lower Greensand ridge is covered by Angular Chert Drift (see p. 127). Economically, the sandy divisions of the Lower Greensand are highly important, the sand of the Folkestone Beds and the rag and chert of the Hythe Beds being exploited at several localities.

The Gault, with an outcrop usually less than a mile wide, occupies the lower part of the scarp-face of the North Downs. It is dug for brickmaking particularly at Dunton Green, Sevenoaks and Borough Green. It is mainly under pasture, but in a few places close to the Chalk outcrops, where a local scarp drift or hillwash covers the Gault, there is some cultivation.

The Upper Greensand crops out between the Chalk and the Gault as a lenticular development in the western half of the district only; to the eastward it is represented by beds in Upper Gault facies; it is but a few feet thick, and is often difficult to detect in the field.

The Chalk covers 31 square miles of country in the north and north-west. Except for the bare outcrops on the scarp-face and on the sides of the coombes, it is largely covered with Clay-with-flints. The outcrop is mainly that of Upper Chalk, while the Middle and Lower subdivisions crop out along the lower part of the scarp-face and in the bottoms of the coombes on the dip-slope.

Superficial deposits include Clay-with-flints and various Head Deposits, among them a number of small patches of angular material derived from the escarpments both in the eastern and western ends of Holmsdale and on the Weald Clay south of the Greensand scarp. The river deposits of the Darent and the Eden are arranged in terraces closely bordering the rivers but the individual gravel patches are generally small. Those of the Medway are, however, more extensively spread over the Weald Clay plain on the eastern side of the district, and consist largely of brickearth, with some gravels.

Within the district two deep boreholes penetrating totally concealed formations have been put down at Penshurst. During 1897–99 a trial bore for coal was drilled to a depth of 1867 ft (Whitaker 1908, p. 231), and in 1938 a trial borehole for oil encountered Palaeozoic Rocks at about 4500 ft below sea level; it passed through Jurassic and Carboniferous strata. Another trial borehole for coal at Old Soar (Whitaker 1908, p. 237), 2½ miles S.E. of Ightham, was put down in 1899 to a depth of 858 ft, but penetrated nothing older than Wealden.

The district still retains many traces of its former iron industry. Straker (1931, pp. 218–81) records some fourteen sites of local bloomeries, furnaces and forges. Of these nine are situated on Wadhurst Clay, two on Ashdown Sand and the others on Weald Clay, Upper Tunbridge Wells Sand, and Brickearth overlying Tunbridge Wells Sand. In addition, there was a site in Devils

Gill some 700 yd west of Knowles Bank near Capel. He also records mine-pits at Crippenden and near Castle Hill, Tonbridge. In both localities the pits are in Wadhurst Clay. Handcock (1910, p. 522) mentions mine-pits in Wadhurst Clay in a wood [543434] between Southborough and Tonbridge, about ¼ mile N.N.W. of Old Forge. Another area retaining the name Minepit Wood lies just north-east of Nightingale Farm [58104405] between Southborough and Tonbridge. An ironstone band near the base of the Wadhurst Clay has been traced at several localities by the bell-pits along its outcrop.

Most of the innumerable pits on the Wadhurst Clay are now grown over or full of water.

An interesting account of the mining is given by Topley (1875, p. 334) who suggests that as well as iron ore, shelly calcareous ironstone was dug as a flux, both occurring near the base of the Wadhurst Clay.

TABLE OF STRATA

The geological formations indicated by colours and symbols on the geological map and sections, (Sheet 287), are as follows:

SUPERFICIAL DEPOSITS (DRIFT)

			Estimated thickness in feet
RECENT	Alluvium	Mud, with sand and gravel	up to 20
	Brickearth	Buff loam	up to 15
	Dry Valley Gravel	Sand and gravel	up to 10
	Terraced River Gravels		up to 20
	Buried Channel Deposits (not shown on map)		up to 40
AND	Head	Unstratified downwashes with angular stones	up to 10
	Clay-with-flints	Loamy clay with pebbles and angular flints	6
PLEISTOCENE	Disturbed Black-heath Beds	Sand with pebbles	—

Estimated
thickness
in feet

SOLID FORMATIONS

CRETACEOUS	Chalk	Upper	White chalk with flints	up to 120
		Middle	White chalk with few flints	180–250
		Lower	Marly grey chalk	190–250
	Upper Greensand		Glauconitic sand and malmstone	0–42
	Gault		Grey clay	225–325
	Lower Greensand	Folkestone Beds	Coarse ferruginous sand	105–225
		Sandgate Beds	Glauconitic clayey sand	2–20
		Hythe Beds	Sand with layers of rag and chert	up to 200
		Atherfield Clay	Grey and buff sandy clay	30–50
	Wealden Beds	Weald Clay	Grey clay and mudstone with sandstone and limestone bands	up to 1100
		Hastings Beds	Upper Tunbridge Wells Sand — Fine buff sand and sandstone	up to 230
			Upper Grinstead Clay — Grey clay	0–25
			Cuckfield Stone — Fine sand and sandstone	0–25
			Lower Grinstead Clay — Grey clay	0–25
			Ardingly Sandstone — Massive sandstone	0–60
			Lower Tunbridge Wells Sand — Fine sand and sandstone	70–100
			Wadhurst Clay — Grey clay with ironstone and mudstone bands	130–180
			Ashdown Beds — Fine sand and sandstone	up to 750

The following formations have been proved in the Penshurst 1938 Borehole:

		Ft
JURASSIC	Purbeck Beds	about 560
	Portland Beds	about 188
	Kimmeridge Clay	about 1455
	Corallian Beds	278
	Oxford Clay and Kellaways Beds	271[1]
	Cornbrash and Great Oolite	about 185[1]
	Inferior Oolite	about 298
	Lias and Rhaetic?	957
	Breccia of Unknown Age	12
CARBONIFEROUS	Carboniferous Limestone Series proved to	958

The Buried Channel Deposits (see p. 125) are everywhere overlain by Alluvium and cannot be portrayed separately on the One-inch geological map.

H.G.D., S.C.A.H., C.R.B.

---

[1]See detailed description for notes on the dips in these formations.

Chapter II

STRUCTURAL GEOLOGY;
CONCEALED FORMATIONS

STRUCTURE

THE CENTRE of the Sevenoaks map area lies to the north of the main axis of the Wealden dome. The beds thus dip generally northwards, the lowest formations being exposed in the south, and the higher members succeeding northwards. West of Tonbridge the regional dip is about two degrees to the north-north-west, and eastwards of this line it is of about the same order to the north-north-east. This simple major structure is complicated by a considerable amount of minor folding and faulting. Two belts of folding are conspicuous, one in the Hastings Beds along the south of the district and one in the Lower Greensand extending from Limpsfield to West Malling. Folds have been noted in the Chalk, and it is probable that others affect the Weald Clay.

Hopkins (1845, pp. 12–13) recorded a line of folding through Bidborough and Brenchley. This fold crosses the area from just north of Dry Hill [432418][1] in the west to just north of Brenchley in the east, and thence eastwards. Along the line of its axis it is responsible for bringing up inliers of Wadhurst Clay beneath the Tunbridge Wells Sand or of Ashdown Beds beneath the Wadhurst Clay, clearly seen on the map at Gilridge [457424], Penshurst, Forest Farm [596439], to the east and west of Gedge's Farm [660430] and $\frac{1}{4}$ mile N. of Brenchley church [681426]. It is better to consider this line of uplift not as one simple fold but a series of smaller anticlines arranged *en echelon*.

The three inliers of Ashdown Beds around Gilridge in the west are exposed on the flanks of a dome which is elongated along a line running from Gilridge to Markbeech and having its structural culmination to the south of Gilridge. Here the base of the Wadhurst Clay lies above 350 ft and from this crest the base falls away in all directions so that it is at 150 ft just east of Cowden [482408] and south-west of Hever [471444], a general dip of about $1\frac{1}{2}°$.

The main Penshurst anticline is a much bigger fold trending east–west but broken by faulting along its northern and eastern sides. The highest structural point is around Ashour Wood [545438] where the base of the Wadhurst Clay is at 300 ft O.D. South of this point the base falls to 125 ft O.D. south of Swaylands [537425], a dip of $3\frac{1}{2}°$ S. On the northern side of the fold, near Redleaf House (now demolished, 524453) the dip is about $4\frac{1}{2}°$ N.W. Strike

---

[1]In this Memoir the positions of localities are indicated by their National Grid reference numbers; all lie within 100-km square 51 or TQ.

The revised One-inch Geological Sheet 287 will be published on the 7th Series Ordnance Survey Topographical Map. Where place names differ from those on the existing published One-inch Geological Sheet 287, both names are incorporated in the text when first mentioned in each chapter. Subsequent references are to the 7th Series names.

9

lines drawn on the base of the Wadhurst Clay show that the nose of the pericline plunges westwards in the region of Wat Stock [508438]. The plunge of the eastern end of the pericline cannot be determined owing to faulting.

In the east the fold to the north of Brenchley is well defined. The main east–west axis runs from just south of Chowning Green [662425] to a point ½ mile N.E. of Brenchley church [685425]. Along this line the base of the Tunbridge Wells Sand lies above 350 ft O.D. falling 6° southwards to 125 ft O.D. east of the church [683417]. Northwards the base falls at the same rate to 150 ft O.D. at a point ⅜ mile N. of the church [68204302]. The small inlier of Ashdown Beds at 681426 lies to the north of the fold axis. The base of the Wadhurst Clay is seen to fall steadily northwards where its junction with the underlying Ashdown Beds can be traced.

South of this line of anticlinal folding a small syncline preserves an outlier of Grinstead Clay, Cuckfield Stone and Upper Tunbridge Wells Sand around Finch Green [508415]. Only the northern limb of the syncline is seen in this map area; the southern limb lies on the adjoining Tunbridge Wells (303) Sheet.

The angle and direction of dip of the various strata are best deduced from their regional relationships. Individual dips of sandstone are usually seen to be distorted by cambering (see p. 77) when measured in natural sections or small man-made openings.

Valley bulging accounts for many of the anomalous dip readings in clays exposed in stream sections and is also liable to occur at the foot of the Lower Greensand and Chalk escarpments (see p. 144).

Faults are the dominant structural element of the Hastings Beds, the majority running approximately east–west parallel to the long axis of the dome and downthrowing to the north. The largest of these faults, extending from near Lingfield in the west to Paddock Wood in the east, forms the boundary between the Hastings Beds and the Weald Clay of the map area. The 30° dips noted in the sandstone in the stream bed north-east of Brook Street [45884406] may be related to this fault or are possibly due to valley bulging. This large boundary fault is dislocated at three points by small north–south faults.

In the railway cutting south of Hever Station [46704435] the fault plane of a north-east–south-west trending fault can be seen throwing the Lower Tunbridge Wells Sand down to the north against the Wadhurst Clay to the south. Farther east in the grounds of Hever Castle this fault is responsible for dips of 45°, 30° and 35° S. in the sandstones exposed on the south side of the fault in the old quarries and ornamental gardens at 47874492, 47944502 and 48004507.

The effect of the north-north-east–south-south-west trending fault which runs from Dry Hill Camp [433415] to just north of Clatfields [442432] can be seen in the old quarry [43824239] south of Greybury where this flaggy sandstone dips at 25° N.W. Some 25 yd W. the beds are horizontal. In the lane and fields immediately to the east the Wadhurst Clay can be proved by augering. Farther north the Ardingly Sandstone is seen as a thin flaggy sandstone folded into a sharp anticline parallel to the fault, in the small stream [44174308] to the south of Clatfields. The beds dip 70° E.S.E. and 5° W.N.W. A small clay-pit [44214309] 40 yd E. marks the outcrop of the Wadhurst Clay. Dips of 15° E.S.E. in the old quarry near Ockhams [44404335] may be associated with this same fault while the dip of 20° N.W. 300 yd to the north-east [44524358] may result from the westerly continuation of the fault seen in the Hever railway cutting.

A series of old quarries at Brooker's Farm [498424] south of Chiddingstone Hoath is in the Ardingly Sandstone. The southern face of these pits is slickensided and is the fault plane of an east-north-east trending fault. This fault is responsible for the thinning of the Lower Tunbridge Wells Sand and Wadhurst Clay in the direction of The Grove [518429]. The fault plane dips steeply north-north-west and a thin veneer of Grinstead Clay has been preserved around the northern, downthrow, side of the most westerly pit [49704233].

A small graben of Wadhurst Clay is preserved in the Ashdown Beds in Hawk's Wood [553441]. Old pits have been dug in the clay of this small faulted inlier at 55254410 and 55454408.

A complex of small faults in the Mabledon Park Hospital vicinity is best known from the Quarry Hill section [586449]. The fault exposed in this quarry (Plate VA) has always been regarded as a reversed fault thrusting Ashdown Beds from the south over the Wadhurst Clay to the north (see p. 44). Mapping carried out in 1965 has shown that the 'Ashdown Beds' are in fact Lower Tunbridge Wells Sand overlying the Wadhurst Clay conformably and down-thrown to the south against the Wadhurst Clay exposed in the quarry. A trial borehole for the bridge foundation for part of the Tonbridge By-pass by the lodge [58304495] to Mabledon Park passed through 11 ft 6 in of soil and down-washed sands into the Tunbridge Wells Sand and continued in the sand to a depth of 57 ft (230 ft O.D.); it then passed into clay for a further 23 ft. The sand/clay junction is interpreted as the fault interface, the Lower Tunbridge Wells Sand being on the south side of the fault and the Wadhurst Clay on the north.

In the railway cutting to the east Prestwich and Morris (1846) noticed a normal fault downthrowing Wadhurst Clay to the south against Ashdown Beds to the north. Although now badly overgrown the position of the fault can still be recognized.

At the western end of this fault complex a small fault is exposed in the road bank at Upper Haysden [56184462] throwing Wadhurst Clay down to the north against Ashdown Beds to the south.

In the Lower Greensand two types of fold are evident. The more conspicuous appears as a discontinuous series of saddlebacks, sometimes *en echelon*, each saddle individually involving narrow strips of ground not more than 200 yd in maximum width. The second type is of gentle flexures of large amplitude.

The limbs of the sharp saddleback folds dip at angles up to 85° and more; usually the northern limb is the more steeply inclined. The Hythe Beds are principally affected, but some folds occur in the Folkestone and Sandgate Beds. In a belt of ground up to a mile in width between Westerham and Sevenoaks this type of folding has repeatedly brought up Hythe Beds, the crests of the folds frequently coinciding with ridges in the ground surface in each of which hard chert that occurs at the top of the Hythe Beds forms a hard core. Conspicuous outliers of Sandgate and Folkestone Beds are preserved in the intervening areas. The Hythe Beds have been quarried at numerous places along these folds, and such is the restricted width of the zones of disturbance that both limbs may be exposed in a single quarry, (see p. 12 and Plate I).

These sharp folds occur in several parallel lines, with others trending obliquely to them. Fitton (1836, pp. 133–6), Hopkins (1845, pp. 21–2) and Topley (1875, p. 234), however, appear to give an impression that they belong to a single line of disturbance. They exert a marked influence on surface drainage and ground-water flow (p. 150 and Fig. 5).

The most westerly of these saddles occurs at Moorhouse Bank, W.S.W. of Westerham [430533] (Topley 1875, p. 234), both northerly and southerly dips being noted in a small pit 150 yd N.W. of Moor House (see p. 68), and southerly dips recorded 400 yd E.N.E. of the house. The fold here coincides with a topographical feature, there being on the north of its axis a steep slope down to the River Darent which continues towards Squerryes Court.

Another fold trends in a similar direction south of Westerham, as shown by southerly-dipping strata behind the pump house [444536] 420 yd N.E. of Squerryes Court and again at the north-east corner of the fish pond [459539] 460 yd E.S.E. of Dunsdale.

The southern limb of a parallel fold farther north was shown by a section in a sunken road from Westerham towards Dunsdale about 250 yd E. of Westerham church, where dips range from 20° to 35° S. This fold plunges westward and shows minor rolls with north to south axes.

Limbs of another fold to the north-east of the last have been exposed in a sunken road 220 yd N. of Valence [45955430], and eastward the fold is further indicated by beds with a southerly dip in a small pit 300 yd N.E. of Valence; east of this the fold gives rise to a prominent ridge which, near the last-named pit, is displaced some 50 to 60 yd N., possibly by a fault.

South of Brasted two folds depart from the east–west trend, one running in a north-easterly direction from 200 yd E. of Vines Gate [46905375] to Colinette Farm, formerly Brasted Place Farm [47505425], where it meets the other, traceable along a north-west axis from a point 300 yd W. of the Hospital, formerly the Workhouse [482537].

A short distance east yet another saddle runs east by north from Manor Farm [494542], south-east of Sundridge, to the railway cutting at Sevenoaks. Both horizontal bedding and southerly dips were noted at Manor Farm and northerly and southerly dips in a pit 400 yd E. of the farm. Similar dips indicated the fold in old pits respectively 450 and 250 yd W.S.W., and 550 yd and 1000 yd E. of Greenlane Farm [507544] and again 300 yd N. of Kipping-ton House, formerly Bishop's Court [52205435]. In the railway cutting changes of dip were noted and the axis of the fold was marked as being 280 yd N.N.W. of the northern end of the tunnel. No trace of this fold was found eastward of the cutting.

About ¾ mile northward a long saddle originally described by Fitton (1836, pp. 133–6) is characterized by steep dips, and runs parallel with the last from about 250 yd S. of Brasted eastwards to Cold Arbor, formerly Coldharbour [505551], where it unites with another trending east-south-east from about 700 yd N. of Sundridge church; the combined fold then continues eastwards to Sevenoaks railway station. In the mile-long portion south of Brasted and Sundridge dips ranging up to 34° to 54° S., and from 34° to 56° N., have been recorded. In 1932 the complete arch of this fold was exposed in a large road-stone quarry at Dryhill [497552] (Frontispiece, Plate I) and dips up to 41° N.

and 85° S. were measured. A full description of this section has been made by Dighton Thomas (Wright and Dighton Thomas 1947, pp. 318–9, plates 24, 25).

About 700 yd N.E. of Sundridge church the more northerly fold showed dips of 46° N. and 30° S; this fold was also traceable in the Dryhill quarry.

Eastwards of Dryhill the anticline is accompanied by a complementary syncline, lying about 60 yd to the south, which becomes weak near Sevenoaks and is not traceable eastwards of the railway there.

Indications of other saddle folds are to be seen in the railway cutting 750 yd S. of Sevenoaks railway station, and near the tunnel entrance, where the fold is accompanied by faulting (Topley 1875, p. 133). West of the railway, dips up to 4° N. and 55° S. have been measured in the first-named of these two folds.

North of Sevenoaks a saddle (south of the former Bradbourne Hall) is traceable eastwards from 250 yd E. of St. Mary's Church, Riverhead, the northern limb having dips up to 30° N. Topley mentions this fold, but mistakenly regarded as Atherfield Clay beds now shown to be of Sandgate age.

In Knole Park anticlinal and synclinal folds are traceable in a number of old quarries. Beds show dips up to 38° N.E., and 67° S.W. Eastward of the junction of these folds the combined saddle is exposed along a line of abandoned quarries.

The second type of fold affecting the Lower Greensand occurs south of the district showing the saddles, as part of a rather flat-topped dome; the presence of which is indicated along the Lower Greensand escarpment, where the base of the Hythe Beds rises from 300 ft O.D. in the extreme west of the sheet to 500 ft just east of Crockham Hill and remains at levels between 500 and 550 ft as far east as River Hill, beyond which it falls to below 200 ft in the east of the map. North of the escarpment the base of the Hythe Beds maintains a relatively flat form as far as an inlier of Atherfield Clay near Crockhamhill Common, at the southern end of which it lies at 500 ft O.D. and also as far as an Atherfield Clay inlier near French Street where the top of that bed lies at over 500 ft O.D. except at the extreme north of the inlier; and again to a third inlier of Atherfield Clay at Brook Place, evidenced by the base of the Hythe Beds lying above 500 ft O.D. around the southern half of the inlier.

Eastward of this last inlier the beds are inclined to the north and it is probable that these three inliers owe their existence in part to this line of flexure. Around the Crockhamhill Common and French Street inliers observed dips were westerly while on the eastern side of the Brook Place inlier they were easterly. Southerly dips are common on the escarpment but mostly they are due to landslip.

Concerning the eastern half of the area, Topley mentions a line of disturbance between Addington and Offham, and H. J. W. Brown (1925, p. 440) described an anticline through Mereworth Woods and a gentle fold, with an amplitude of about a mile, trending east-south-east through a wood (Birchett Wood) ½ mile N.E. of Roydon Hall [666517]; this crosses the southern half of the outlier of Hythe Beds between Mereworth Castle and Roydon Hall which has, however, suffered much superficial disturbance (see p. 77). A slight arching of the lowest beds of the Hythe Beds south of Adam's Well [656527], ½ mile W. of West Peckham, may be a westwards extension of this fold.

The strata rise again north of this fold in another gentle parallel anticline which rises from Old Soar in the Shode Valley to Broom's Croft [683541] on the eastern margin of the area. C.R.B., S.B., H.G.D.

Of two sharp anticlinal saddle-folds in the Hythe Beds north of Offham the more northerly one appears to correspond with that noted by Topley referred to above. Whereas Topley (1875, p. 234; Plate II) describes the fold as extending eastwards to Leybourne or farther, present evidence indicates rather that it is a sharp fold, confined to the Hythe Beds outcrop, which does not persist more than ½ mile E. of the point [66105857] ⅓ mile N. of Offham church where it is first seen.

The more southerly fold was traced east-south-eastwards from a point 300 yd N.W. of Offham church to near Fartherwell [66755760]. It is marked by a series of hollows, described by Bennett (1908a, p. 284) as being swallow-holes along a joint line of weakness. About 200 yd N.N.E. of Fartherwell dips of about 15° S.W. and 10° to 30° N.E. were seen. This very sharp anticlinal fold may not persist at depth, but it is possible that it continues farther to the south-east, and a dip of 6° S.E. seen ¼ mile S.W. of West Malling church may be on its north-eastern limb. Exposures in the same valley farther west near Offham show only approximately horizontal Hythe Beds until Comp Farm, formerly Comp [645571], is reached and here the dips are to the north and north-east only.

In the north-western part of the map area the Chalk dips north between 1° and 2° with slight rolling and little apparent faulting. Thus at Titsey the base of the Chalk is just over 500 ft O.D. falling to 200 ft O.D. at Otford in the Darent Gap, while the base of the Upper Chalk, lying at 840 ft O.D. at Botley Hill, drops to 650 ft north-west of Trottiscliffe. In the Darent valley the inclination is about 100 ft in ¾ mile.

H. J. W. Brown (1925, p. 446) pointed out a steepening of the dip along the Chalk escarpment between Dunton Green and Oxted, just west of the present area. Among northerly dips recorded in the present survey are 8° above Westerham, 5° near Titsey and inferred amounts of from 3° to 7° near Tatsfield. The dip probably falls again in the Lower Chalk immediately south of the escarpment.

An east-north-easterly fall in level of the crest of the Chalk escarpment which commences in the west of the present area has been noted by Bury (1910, pp. 652, 659; fig. 2, p. 642) and by Wooldridge (1927, p. 71) and it has been regarded as evidence for a warping transversely to the strike which the late Professor Wooldridge concluded to be of pre-Pliocene age.

All the deeper valleys of the Chalk dip slope have been cut down into Middle Chalk, and on the eastern side of the valley west of Biggin Hill the Middle–Upper Chalk boundary rises northwards from near Norheads [409591] to well above the 500 ft contour, in response to a gentle anticlinal roll (p. 107) first noted a little west of Biggin Hill, but perhaps also present farther west-south-west, and attaining a more distinct development eastwards on the Biggin Hill side of the valley. The fold extends east-by-north for ½ mile and then swings gently round to east-by-south towards the Cudham valley. At a point 500 yd N.E. of Costains Farm [422595] beds estimated at 30 to 40 ft above the base of the Upper Chalk lie at about 580 ft O.D., i.e. near the crest of the fold, whereas 500 yd S.S.E. of Single Street [436599] they are at 540 ft O.D.; hereabouts the northerly fall of the Upper Chalk base-line in a short distance becomes accentuated. From this point the flexure trends east-south-east.

The base of the Upper Chalk lies a little above 700 ft O.D. near Brastedhill Farm [463573], north-west of Sundridge, and continues at that level north-eastwards past Sundridge Hill; it does not drop to below the 600 ft level until about ½ mile N.E. of Old Star House, formerly The Beacon [494585]. The base of the Middle Chalk similarly remains at a little above 500 ft O.D. from the scarp below Brastedhill Farm, and around the deep coombe north-west of Chevening, as far as Old Star House, beyond which it falls in conformity east-north-east.

The axis of a very slight syncline runs near Hogtrough Hill [460566], while south of Tatsfield a gentle anticlinal roll is indicated by the rise of the Middle Chalk base-line along the escarpment and also of the outcrop of the Upper Greensand.   S.C.A.H.

CONCEALED FORMATIONS

Two deep boreholes in the Sevenoaks map area provide information on rocks not exposed at the surface.

The earlier boring, put down in 1897–9 in search of coal, was at a site about 1⅝ miles N.E. of Penshurst [54804525]. Surface level is about 90 ft above O.D. A summarized account of the boring was published by Dawkins (1905, p. 30) and a detailed log, including references to some of the fossils, was contributed by him to the Survey Memoir "The Water Supply of Kent" (Whitaker 1908, pp. 231–4). The geological results were described more fully by Lamplugh and Kitchin (1911, pp. 66–77, 183–97; see also Lamplugh, Kitchin and Pringle 1923). This borehole is referred to here as the Penshurst 1899 Borehole.

The later borehole was made in 1938 in search of oil, at a site [542443] 1 mile E.N.E. of Penshurst. The driller's depths were measured from the top of the rotary table at 111 ft above O.D., but the ground-surface level is calculated at 99 ft 6 in; 12 ft have been deducted from the logged depths to arrive at the approximately true depths quoted below. The thickness of the formations has not been corrected for dip as information on this is very limited. Most of the cores show only a low dip, except between 3126 and 3320 ft where dips of 20–45° are recorded. The descriptive log compiled by the oil company's geologists and notes provided by the late F. H. Edmunds have been used in preparing the following account. This boring is referred to here as the Penshurst 1938 Borehole.

The Penshurst 1899 Borehole stopped in Kimmeridge Clay at a total depth of 1867 ft. The Penshurst 1938 Borehole proved a complete sequence of Jurassic rocks resting at 4618 ft on a breccia of unknown age resting on Carboniferous Limestone, which was drilled to 5588 ft.

In the following account Dr. F. W. Anderson and Dr. R. Casey are responsible for the Wealden and Purbeck Beds, and Dr. Casey for the Portland Beds. Mesozoic formations below the Portland Beds have been dealt with by Mr. R. V. Melville and Dr. H. C. Ivimey-Cook, using ammonite determinations by Dr. J. C. W. Cope and Dr. J. H. Callomon. The Carboniferous Limestone is described by Sir James Stubblefield, F.R.S. and Mr. M. Mitchell. The former has identified all the fossils with the exception of the corals, the late S. Smith's identifications of which have been revised by Mr. Mitchell. Petrological notes by Dr. P. A. Sabine have been included in the account.   S.C.A.H.

Wealden (Hastings Beds) at least 552 ft. Both the 1899 and 1938 borings were started in the Ashdown Sand division of the Hastings Beds.

Fine-grained sandstone with bands of silty and sandy shales and a few thin beds of mottled red and grey clay were first passed through (Penshurst 1899, 0–225 ft). Below this, fine-grained silts and thin clay predominate and pass downwards without any marked lithological change into beds of Purbeck age. No fossils were recovered from the upper part of the Ashdown Beds but the more argillaceous beds towards the base contained beds with plant debris, ostracods, fish scales and freshwater mollusca. Plant remains were recorded from the 1899 Borehole at 359, 474 (fronds of *Leckenbya valdensis* Seward), 476 and 506 ft. Ostracods occurred in the same boring at 359, 508, 510, 512, 520 and 536 ft. They are all forms characteristic of the lowest Wealden beds elsewhere within the Weald, i.e. *Cypridea brevirostra* Martin, *C. aculeata* Jones, *C. laevigata* (Dunker), with *Darwinula leguminella* (Forbes), *D. oblonga* (Roemer) and *Rhinocypris jurassica* (Martin).

The phyllopod *Bairdestheria subquadrata* (J. de C. Sowerby) was found at 533 ft. Mollusca were represented by species of *Viviparus*, *Neomiodon* and *Unio*, including *U. gualterii* J. de C. Sowerby, found between 510 and 540 ft.

A whitish sandstone, 7 ft thick, with silty and carbonaceous bedding-planes, becoming shaly below and with a ½-in band of hard calcareous sandstone at the base, is taken to mark the base of the Wealden, at 552 ft, in the Penshurst 1899 Borehole.

Purbeck Beds 562 ft.

Upper Purbeck (*Cypridea setina* Zone)[1] about 128 ft. At Penshurst 1899 this zone extends apparently from 552 to 680 ft, and at Penshurst 1938 from about 425 to 553 ft (estimated). The beds comprise alternations of thinly bedded dark silty shale; clayey shale with brown ironstone bands and nodules; greenish and grey shale; and thin bands of hard calcareous sandstone. Only one sample was seen from the more recent boring, i.e. from between 530 and 560 ft, but this contained the characteristic zone fossil, *Cypridea setina* (Anderson).

In the Penshurst 1899 Borehole, ostracods were found at 590, 628, 629, 634 and 662 ft and included *Bisulcocypris striata* (Martin), *Cypridea setina*, *C. wicheri* Wolburg, *C. menevensis* (Anderson), *C. propunctata* Sylvester-Bradley, *Darwinula leguminella* and *D. oblonga*. Plant remains, including *Equisetites sp.*, were found at 585 and 634 ft. Between 585 and 680 ft species of *Neomiodon* comparable with *N. medius* (J. de C. Sowerby) and *N. elongatus* (J. de C. Sowerby), together with other freshwater mollusca (*Unio* and *Viviparus*), were not uncommon.

Middle Purbeck (*Cypridea vidrana* Zone) about 110 ft. This zone extends from 680 to 790 ft at Penshurst 1899 and from about 553 to 663 ft (estimated) at Penshurst 1938. It comprises a series of dark blue and greenish grey shales with silty intercalations, pyritous and clay ironstone bands, thin layers of impure shelly limestone and calcareous sandstones. The beds are more fossiliferous than those above and contain faunas of both fresh and brackish water facies. Little material was available from this part of the succession in the Penshurst

---

[1]Zonal scheme proposed by Anderson (1940), modified in the light of more recent information.

1938 Borehole, though one of the species characteristic of this zone, *C. inaequalis* Wolburg, was found. In the Penshurst 1899 Borehole increased salinity is reflected in the molluscan fauna, which includes the gastropods *Hydrobia chopardiana* (de Loriol), *Pachychilus attenuatus* (J. de C. Sowerby) and *Ptychostylus harpaeformis* (Koch and Dunker) and the bivalves *Corbula sp.*, *Eocallista* (*Hemicorbicula*) *parva* (J. de C. Sowerby), *Modiolus sp.* and *Neomiodon medius*.

The ostracods recovered from the 1899 Borehole, at depths of 687, 689, 718, 722, 749 and 764 ft, are mostly *Cypridea inaequalis*, but *C.* cf. *menevensis* is common, together with the long-ranging forms *Darwinula leguminella*, *D. oblonga* and *Rhinocypris jurassica*. Species of more saline habitat, such as *Stenestroemia fragilis* (Martin), were occasionally found, and between 785 and 790 ft this species was accompanied by *Bisulcocypris striata* and *Fabanella boloniensis* (Jones). This horizon appears to be the equivalent of the Scallop Bed of the Dorsetshire Purbeck succession.

Middle Purbeck (*Cypridea granulosa* Zone) about 115 ft. At Penshurst 1899 this zone extends from 790 to 905 ft, and at Penshurst 1938 from about 663 to 778 ft (estimated).

The sediments are hard dark shales and silty shales with intercalations of greenish clay, pyritous, hard marly bands and calcareous sandstones, and calcareous shales with nodular limestones. This zone includes the main marine-brackish horizon of the Purbeck Beds, which in the Dorset coast sections is represented by the Cinder Beds, 8 ft 6 in thick and composed almost entirely of the shells of *Liostrea distorta* (J. de C. Sowerby) in a shaly matrix. In the Weald this horizon is not so distinctive but is clearly recognized by its fauna.

No fossils were recovered from the upper part of this zone in the Penshurst 1938 Borehole, but in the 1899 Borehole strata from about 810 to 830 ft contained *Liostrea distorta*, *Myrene fittoni* Casey, *Protocardia major* (J. de C. Sowerby) and other bivalves typical of the Cinder Beds horizon. Below, in the lower half of the zone, marine to brackish beds with such bivalves as *Corbula inflexa* Roemer, '*Gervillella*' cf. *arenaria* (Roemer), *Liostrea sp.*, the minute gastropod *Hydrobia forbesi* Arkell, and ostracods such as *Bisulcocypris striata*, *Fabanella boloniensis* and *Scabriculocypris trapezoides* Anderson alternate with brackish to fresh-water deposits with *Neomiodon sp.*, *Viviparus sp.*, *Cypridea posticalis* Jones and *C. granulosa* (J. de C. Sowerby).

The lowest of these brackish marine bands, i.e. that between 902 ft 6 in and 905 ft in Penshurst 1899, is taken to be the equivalent of those with a similar fauna which mark the base of the Middle Purbeck Beds in Dorset, i.e. the base of the Marly Freshwater Beds.

Lower Purbeck (*Cypridea dunkeri* Zone, upper) about 88 ft. Strata between 905 and 993 ft in the Penshurst 1899 Borehole and between about 778 and 866 ft in the Penshurst 1938 Borehole may be allocated to the upper part of this zone. They comprise black, brownish and pale blue splintery calcareous shales and flaggy cementstones with seams of brown bituminous shale and bands of black limestone. Seams and streaks of gypsum also occur.

No fossils were recovered from either boring but an oolite with *Viviparus* and *Cypridea* was recorded by Boyd Dawkins near the base of the zone in the earlier boring. A band of grey, fine-grained limestone composed largely of ostracod carapaces and 1 ft 6 in thick, recorded from 993 ft, is taken to be the equivalent of the Hard Cockle limestone of the western sequence and to mark the base of the upper division of the zone in the Penshurst 1899 Borehole.

Lower Purbeck (*Cypridea dunkeri* Zone, lower) about 121 ft. The lowest beds of the Purbeck comprise sediments similar to those in the zone above except that gypsum beds are more prevalent. In fact the lower half of this subdivision consists almost entirely of the gypsum beds which constitute the main economic resources of the Purbeck Beds. Macrofossils are rare and poorly preserved but in both borings there were ostracod beds with the usual fauna consisting of *F. ansata* Jones, *F. boloniensis*, *Mantelliana purbeckensis* (Forbes) and *Scabriculocypris trapezoides*.

In both borings the main gypsum beds were about 52 ft thick. In Penshurst 1899 the uppermost Portland at 1114 ft 4 in, a grey calcareous sandstone with trigoniids, was overlain by a thin gypsiferous cementstone. In Penshurst 1938 the top of the Portland Beds was described as a grey sandy limestone at 986 ft.   F.W.A.

Portland Beds (Penshurst 1899, 1114 ft 4 in to 1302 ft; 187 ft 8 in: Penshurst 1938, 986 ft to about 1174 ft; about 188 ft).

About 188 ft of beds are allocated to the Portland Beds. They contain marine bivalves and rare ammonites and are separated fairly sharply from the gypsiferous beds of the Lower Purbeck.

The log of Penshurst 1899 described 131 ft of calcareous sandstone and sandy mudstone overlying more argillaceous beds, which pass down into the Kimmeridge Clay. The upper portion yielded bivalves including *Camptonectes* (*Camptochlamys*) *lamellosus* (J. Sowerby), *Exogyra nana* (J. Sowerby) and *Isognomon listeri* (Fleming). The lower 50 ft of more argillaceous beds yielded *Epivirgatites* cf. *vulgaris* (Spath) and *Progalbanites sp.* indicating the *albani* Zone of the Lower Portlandian at 1301 ft. Other ammonites from these beds include pavloviid fragments from 1255 ft and 1295 ft and indeterminate fragments from 1118 ft and about 1160 ft.

In Penshurst 1938 most of the Portland Beds were cored. Specimens consist of greyish green and grey, calcareous, fine-grained, often silty sandstone with thinner bands of grey sandy limestone. Scattered fossils occur throughout with some pieces of lignite. *Camptonectes* (*Camptochlamys*) *lamellosus* was recorded from a grey sandy and gypsiferous limestone at 986–7 ft together with *Protocardia dissimilis* (J. de C. Sowerby); other bivalves found in these sandy beds include *Exogyra*, *Gryphaea?*, *Isognomon* and a trigoniid. A fragment of a large pavloviid ammonite occurred at 1084 ft, a perisphinctid at 1119 ft and *Crendonites sp.* at 1128 ft. Cores taken between 1103 and 1136 ft were of silty shales, locally slightly bituminous. No core was taken between 1136 and 1751 ft, but cuttings from the higher parts of this sequence were described as fine-grained bluish black calcareous shale.

As Penshurst 1899 has evidence of the *albani* Zone in the argillaceous beds at 1301 ft a similar thickness of shaly beds has been included in the Portland Beds of Penshurst 1938.

Kimmeridge Clay (Penshurst 1938, about 1455 ft.)

Upper Kimmeridge Clay (Penshurst 1899, 1302 to 1867 ft (565 ft to base of hole): Penshurst 1938, about 1174 to 1771 ft+; 597 ft+). Penshurst 1899 was completed at a depth of 1867 ft in beds attributed to the Upper Kimmeridge Clay. This thickness included beds with *Pavlovia spp.* between 1320 and 1471 ft, *Pectinatites* (*Virgatosphinctoides?*) *sp.* at 1542 ft and *Pectinatites sp.* at 1864 ft. The fauna also includes *Discina latissima* (J. Sowerby) at 1541 ft and *Lingula ovalis* J. Sowerby between 1317 and 1665 ft. The bivalves included *Grammatodon* cf. *rhomboidalis* (Contejean) between 1361 and 1458 ft; *Isocyprina* (*Venericyprina*) *argillacea* Casey between 1320 and 1394 ft; *Modiolus* (*Musculus*) *autissiodorensis* (Cotteau) between 1565 and 1661 ft; *Hartwellia hartwellensis* (J. de C. Sowerby) between 1657 and 1771 ft; and *Protocardia morinica* (de Loriol) and *Lucina miniscula* Blake throughout. The pyritized remains of the crinoid *Saccocoma* were found between 1794 and 1796 ft (Bather 1911). *Saccocoma* is best known from horizons equivalent to the Blackstone of Dorset, but in the Warlingham Borehole it was recorded both in beds which may correlate with this horizon (i.e. the top of the *wheatleyensis* Zone) and from 200 ft lower in beds corresponding to the *eudoxus* Zone (Casey 1958, p. 48).

These observations suggest that the Upper Kimmeridge Clay at Penshurst has a minimum thickness of 597 ft compared with only 371 ft 10 in at Warlingham. The precise thickness of these beds at Penshurst is uncertain owing to the indefinite junction with the Portland Beds and the lack of core samples from between 1771 and 2229 ft.

Lower Kimmeridge Clay. Core taken between 2229 and 2250 ft yielded *Aulacostephanus sp.* at 2230 and 2240 ft from grey slightly calcareous silty shale and mudstone, indicating the Lower Kimmeridge Clay possibly of the *eudoxus* Zone. The log records fine-grained grey and bluish black calcareous shales from 1771 ft to the top of this core, with more sandy shale bands in the lower part of the sequence. Below 2250 ft the record shows increasing amounts of pale sandy shale in the darker calcareous shales. Further core was taken from 2296 to 2333 ft and contained grey calcareous siltstones and limestones.

The occurrence of *Rasenia* (*Rasenioides*) cf. *lepidula* (Oppel) in grey calcareous shales between 2316 and 2397 ft, associated with *Rasenia* (*R.*) *thermarum* (Oppel) between 2395 ft 6 in and 2397 ft, *Rasenia* aff. *eulepida* Schneid, *Rasenia* aff. *moeschi* (Oppel) and *Aulacostephanus* cf. *mutabilis* (J. de C. Sowerby) from 2397–8 ft[1] indicates the *mutabilis* Zone.

No further cores were taken between 2398 and 2417 ft but thereafter fairly numerous specimens are available to a depth of 2616 ft. These show a variety of mainly sandy facies down to about 2494 ft and thereafter silty and calcareous mudstones predominate. These beds are regarded as of Lower Kimmeridgian age on the basis of a specimen of *Prorasenia* cf. *bowerbanki* Spath at 2609–10 ft which indicates the *Pictonia baylei* Zone; other ammonites include *Ringsteadia?* or *Pictonia* from 2583–91 ft, an indeterminate *Prorasenia?* at 2616 ft and two indeterminate perisphinctids at 2568–72 ft and 2609–29 ft. *Pinna sp.* is common between 2568 and 2609 ft. The arenaceous beds between

---

[1]The use of a hyphenated depth indicates that no exact evidence of depth is available within that range.

c

2424 and 2494 ft contain an upper group with grey bioturbated slightly calcareous mud and silt, originally coarsely interlaminated with fine whiter sandstone, passing down into coarser more massive, often greenish grey, sandstones with calcite in the matrix and very sandy limestones with occasional echinoderm and shell fragments. Pale grey, shell-fragmental limestones often rich in quartz, ooliths and pyrite are interbedded with greyish brown calcareous sandstones below 2474 ft. Bioturbated grey calcareous sandstones and dark grey silty mudstones at 2492–4 ft overlie a sequence of finer grained grey silty mudstones, with marlstone beds at 2512 and 2520 ft, down to 2544 ft. Between 2544 ft 6 in and 2546 ft occurs a grey calcite mudstone with large alga-coated pisoliths and *Exogyra*. Dark grey calcareous mudstones and siltstones with fragments of *Pinna*, *Chlamys* and *Gryphaea*, with pale grey calcite mudstones at about 2556 and 2597 ft, overlie grey mudstones which at 2629–49 ft contain *Amoeboceras* (*Prionodoceras*) *sp.* in a facies reminiscent of the Ringstead Waxy Clays of the Corallian Beds.

Thus the total thickness of the Kimmeridge Clay in this borehole, including the arenaceous beds near the base, is estimated to be about 1455 ft.

Corallian Beds (2629 to 2907 ft; 278 ft).

The *Amoeboceras* (*Prionodoceras*) *sp.* and *Decipia sp.* from 2629–49 ft suggest the *Ringsteadia pseudocordata* or *Decipia decipiens* zones of the top of the Corallian Beds (Upper Oxfordian) and they are found in grey mudstones locally rather bituminous and with scattered bivalves. The record indicates fine-grained green to black calcareous shales with streaks of hard bluish grey marl between 2649 and 2731 ft and then further shales with bands of grey sandy limestone to the next core at 2759 ft. Below this depth grey calcareous siltstones and silty limestones with bands of mudstone continue to 2804 ft. Between 2806 and 2813 ft numerous specimens of *Cardioceras* (*Cawtoniceras*) *sp.* occur in grey calcareous silty mudstone, including *C.(C.) cawtonense* (Blake and Hudleston) at 2808 ft, indicating the upper part of the *plicatilis* Zone (probably the *antecedens* Subzone). *Lingula craneae* (Davidson) was found at 2808 ft.

The lower part of the Corallian Beds is formed of a mass of limestones approximately 58 ft thick. The pale grey limestones contain occasional radioles of *Paracidaris florigemma* (Phillips) at 2844 ft, echinoderm fragments, traces of coral at 2848 ft, terebratulid, rhynchonellid and bivalve fragments. Below 2853 ft the limestone is often crystalline and detrital, ooliths and pisoliths occur between 2857 and 2859 ft and a pseudobrecciated? calcite mudstone at 2856 ft. An incompletely cored layer of *Rhaxella* chert occurred at 2859 ft, apparently the first record of this rock type in the Corallian Beds of the Weald. The base of the Corallian Beds has been taken at a grey calcareous mudstone with a pitted surface encrusted with *Lopha* at 2907 ft.

Oxford Clay and Kellaways Beds (2907 to 3178 ft; 271 ft).

Ammonites indicating the *bukowskii* Subzone of the *cordatum* Zone occur in grey calcareous mudstone at 2916 ft, with *Cardioceras* (*Scarburgiceras*) aff. *praecordatum* (Douvillé) associated with *Grossouvria sp.* within 2912–2912·9 ft, *Hecticoceras sp.* and *Vertebriceras sp.* at 2918 ft. The *Vertebriceras* may indicate the highest part of the underlying *mariae* Zone, *praecordatum* Subzone. The lower subzone of this zone, that of *C.(S.) scarburgense*, is proved at 2976 ft and on to about 3028 ft with a rich ammonite fauna including *C.(S.)* *scarburgense* (Young & Bird) at various depths between 2976 and 3035 ft; *Cardioceras sp.* between 2955 and 2986 ft; *Creniceras renggeri* (Oppel) at 2998–3005 ft and its supposed sexual dimorph *Taramelliceras richei* (de Loriol) at 2994–8 and 3018–28 ft (Palframan 1966); *T. episcopale* (de Loriol) at 3028 ft; *Quenstedtoceras (Q.)* cf. *mariae* (d'Orbigny) at 2989 to 2991 ft; and *Hecticoceras* cf. *matheyi* de Loriol at 2995–3005 ft.

No core was taken between 3035 and 3114 ft or between 3134 and 3158 ft, both intervals being recorded as in fine greyish black shales with greyish black calcareous shales. *Binatisphinctes?* occurs at about 3114 ft and may indicate the basal part of the *athleta* Zone; greyish brown calcareous mudstones with bivalves continue to 3127 ft; below 3118 ft they contain a fauna with *Kosmoceras* and other ammonites which indicate the *coronatum* Zone. In this fauna *Erymnoceras sp.* occurs at 3121 and 3127 ft; *Hecticoceras brighti* (Pratt) at 3119 and 3124 ft 6 in; *H. sp.* at 3127 ft, *Kosmoceras castor* (Reinecke) at 3119 ft, *K. gulielmi posterior* Brinkmann at 3124 ft; *K. pollucinum* Teisseyre at 3127 ft; *K. sp.* at 3124 ft 6 in to 3127 ft and *Reineckeia sp.* at 3118 ft.

The succession is broken by a shear plane dipping at 20° and filled with calcite at a depth of 3128 ft. The remainder of this core sequence (3128–34 ft) is in slightly calcareous greyish brown mudstone with *Meleagrinella*; it is of Oxford Clay type but without ammonites. A dip of 20° was recorded at 3128 ft. These mudstones are recorded as continuing to about 3158 ft and the next core (3160–78 ft) is in greyish brown fine-grained sandstone, silt and mudstone, bioturbated and locally calcareous, and representing the Kellaways Beds. Obscure plant fragments occur in these sandstones and also oysters resembling *Liostrea (Catinula) alimena* (d'Orbigny) between 3161–7 ft, though as they are preserved as internal moulds no ribbing was visible. The lowest specimen of these beds was of hard massive, non-calcareous sandstone at 3177 ft 6 in and the contact with the underlying beds is not preserved.

Cornbrash? and Great Oolite (3178 to about 3363 ft; about 185 ft).

Grey crystalline limestones with *Entolium* occur between 3178 to 3180 ft above a sequence of grey muddy limestones and grey calcite mudstones with scattered ooliths, shell detritus and calcite veining down to 3189 ft. The brachial valve of a terebratuloid with a long median septum at 3183 ft may be referred either to *Digonella* or to *Microthyridina* so that the identity of the highest limestones with the Cornbrash is not firmly established. *Kallirhynchia* is present both above and below this specimen at 3180 and 3184 ft. The calcite mudstones with ooliths and shell detritus recall the Forest Marble in lithology and they contain bands of calcareous mudstone. The bivalve fauna includes *Entolium* from 3180 to 3182 ft 6 in and *Camptonectes* and *Pseudolimea* at 3184 ft. The specimens from 3180 ft show slight shear marks similar to but less marked than those at 3128 ft.

Limestones of the Great Oolite Series are found to a depth of 3338 ft and include a sequence of calcareous shales and limestones between 3278 ft and 3308 ft. The limestones between 3182 and 3308 ft have recorded dips of between 22° and 45°. No correction has been made to the thickness of the formation on account of these dips. The higher limestone group is dominantly of grey crystalline limestone with shell detritus which includes rhynchonellid, terebratulid and echinoderm fragments, and calcite filled vugs. It becomes paler and more oolitic below about 3221 ft. A one-inch band of sheared calcite occurs at 3252 ft and the grey detrital limestones from this level to about 3258 ft are much veined with calcite and there are several levels with minor shear faces. Between 3258 and 3268 ft is a grey fine-grained oolite overlying more crystalline limestones down to 3284 ft 6 in. The limestones contain *Kallirhynchia* at 3191 ft 6 in and 3257 ft; *Camptonectes* (*Camptochlamys*) at 3257 ft; *Liostrea* or *Exogyra* at 3257 and 3264 ft; and *Placunopsis socialis?* Morris and Lycett at 3191 ft 6 in.

A bed of grey calcite mudstone with *Entolium corneolum* (Young & Bird) and *Epithyris* occurs between 3284 ft 6 in and 3286 ft and calcareous mudstones with occasional calcite mudstones and grey limestones down to 3320 ft 6 in. A specimen of *Gervillella* was found at 3319 ft. Grey crystalline limestones with echinoderm and coral fragments occur between 3325 and 3326 ft and succeed calcite mudstones, with *Rhynchonelloidella* at 3322 ft, down to 3334 ft. Grey calcareous mudstone, with shear planes, found at 3336–7 ft and between 3348 and 3352 ft, contains a fauna including *Protocardia* and *Lingula*, which continues to the base of this section of core at 3352 ft. This mudstone sequence may be tentatively correlated with the Lower Fuller's Earth.

Inferior Oolite (about 3363 to 3661 ft; about 298 ft).

The Inferior Oolite may be represented by the unfossiliferous oolites recorded between 3363 and about 3661 ft. The core recovered from between 3365 and 3411 ft is of pale grey oolite with shell and echinoderm fragments, some grey detrital limestones and clay partings. The only other core was from between 3544 and 3584 ft and consisted of grey shell-detrital limestones with ooliths, echinoderm fragments and rather more mud. The log records that the upper group of oolites extends to 3498 ft, though becoming more shaly in the lower part and with bluish grey calcareous shale containing streaks of grey oolitic limestone between this depth and about 3558 ft. Between 3562 and 3565 ft the limestones contain alga-coated pisoliths and large ooliths.

Upper Lias (3661 to about 4008 ft; about 347 ft).

Cores, dominantly of oolitic and sandy ironstone, were taken between 3661 and 3762 ft, at 3784 to 3801 ft, and 3921 to 3941 ft. The highest beds attributed to the Upper Lias are hard greyish brown sandstones and calcareous siltstones with clay intercalations and mica flakes, extending down to 3669 ft where they overlie limonitized calcareous siltstones. These dark reddish brown ferruginous beds continue down to 3699 ft and contain calcareous limonite oolite at 3677 ft, limonite siltstone at 3686 ft, and ferruginous limestone and siltstone at 3695 and 3699 ft respectively. A specimen of *Pleydellia?* at 3677 ft suggests the highest subzone of the *levesquei* Zone; other fossils include *Lingula* at 3665 ft; *Gryphaea?* at 3672 ft; *Meleagrinella?* at 3677 ft and *Entolium liasianum* (Nyst) at 3682 ft. Between 3700 and 3707 ft are grey micaceous sandstones and siltstones, the lower part of which is bioturbated and more ferruginous. A further group of ironstones contains *Grammoceras thouarsense* (d'Orbigny) at 3707 ft indicating the zone with that index fossil; *Grammoceras sp.* was found at 3708 ft and between 3714–32 ft, *Ctenostreon rugosum* (W. Smith) at 3717 ft, and *Gryphaea*, *Entolium* and *Meleagrinella* between 3696 and 3720 ft. The ironstones of this group include calcareous limonite oolite at 3729 ft and limonite oolite at 3732 and 3736 ft. The ironstone lithologies show heavy oxidation and the sequence is regarded by Taylor (*in* Hallimond and others 1951, 78–9) as also showing rhythmic sedimentation. The depths quoted there are not corrected for the height of the drilling platform. They are found down to about 3740 ft interbedded with dark grey, often bioturbated, micaceous sandstones and siltstones. Specimens from between 3740 and 3762 ft are of grey micaceous siltstone with thin shell-detrital limestones containing echinoderm debris and shell fragments with *Entolium* and *Liostrea* at 3757 ft. Similar dark grey micaceous siltstones are found again between 3786 and 3799 ft. The record shows further grey slightly micaceous silty shales with some bluish grey argillaceous limestone and laminated calcareous shale down to the next core taken between 3921 and 3941 ft. This core included grey silty limestones and dark grey laminated silts and clays, often very contorted. The sequence below, to the next core at 4040 ft, is recorded as showing more grey argillaceous and sandy limestones than in the higher beds and then a change to hard white calcareous gritty and ferruginous beds below 4020 ft.

Middle and Lower Lias, Rhaetic? (about 4008 to 4618 ft; about 610 ft).

Shell-fragmental limestones with dark ferruginous ooliths and bands of dark reddish grey silty ironstone are found within 4040 to 4060 ft. They contain a fauna with *Entolium*, *Liostrea*, a small gryphaeate oyster, *Meleagrinella substriata* (Zieten), *Pseudopecten*, *Tetrarhynchia* and *Lobothyris?*. The fauna and lithology suggest that these beds may belong to the Middle Lias.

Only one further core was taken in Mesozoic rocks in this borehole— between 4538 and 4541 ft. It contained grey compact locally silty limestone with recrystallized shell fragments which may have been of *Liostrea*. These beds are referred to the Lower Lias. The strata between 4060 and 4538 ft are recorded as hard white calcareous grit with micaceous silt partings down to 4098 ft, a sequence of slightly calcareous shales down to 4178 ft and then similar shales with streaks of fine-grained light grey argillaceous limestones becoming increasingly sandy downwards to 4423 ft. Below this there is an indication of a similar sequence with calcareous shales to about 4443 ft, shales and limestones to about 4503 ft and more sandy beds to about 4552 ft. Sixty feet of pale grey and whitish brown crystalline limestone with streaks of black slightly silty calcareous shales were recorded between 4558 and 4618 ft. No positive evidence has been obtained for the age of these limestones.

'Lower Carboniferous Breccia' (4618 to 4630 ft; 12 ft)

The log records 12 ft of "limestone with yellow marly clay and grey sandy marl with red stain". This bed may be regarded as the weathered top of the Carboniferous Limestone. However in the Henfield Borehole Chaloner (1962) showed that there the thick sequence of brightly coloured marls and clays with breccias, conglomerates and limestones may be correlated with the Rhaeto-Lias between 4890 and 4997 ft and possibly with the Trias between 4997 and 5060 ft. No evidence for the age of these beds has been obtained from the cuttings of the Penshurst Borehole, and the breccia is therefore given no age designation in the generalized section on the map.   R.V.M., H.C.I.-C.

Carboniferous Limestone (4630 to 5588 ft; 958 ft)

Core samples totalling 175 ft in length were taken from thirteen horizons at intervals within the Carboniferous Limestone, but recovery was poor and averaged less than 25 per cent, so that detailed depths within a core length cannot be stated. The uppermost core is from immediately below the 'Lower Carboniferous Breccia' of the oil company's graphic log and the base of the lowest core is from a depth of 5441 ft. From the evidence of the chippings, Carboniferous Limestone continued to the base of the borehole at 5588 ft below the surface. The greater part of the cores consists of pale greyish brown, fine-grained massive limestone with negligible dip, and the rocks are of Viséan age. Cherts are present in the topmost core and breccias at the lower depths.

The first core length is from 4630 to 4640 ft (recovery 30 per cent). The uppermost rock type recovered is a pale grey to white chert showing, in section, foraminifera, echinoderm spines, trepostomatous bryozoa, Productid spines and a turreted gastropod. Below this occurs pale greyish brown fine-grained limestone with *Calcisphaera sp.*, *Koninckopora inflata* (de Koninck), *Lithostrotion martini* Milne Edwards and Haime [S$_2$ form of Whittard and Smith 1943, pl. 15, fig. 4], *Syringopora sp.*, *Athyris expansa* (Phillips), *Dielasma sp.*, *Megachonetes* aff. *papilionaceus* (Phillips), Orthotetid and Productid fragments.

The second core was taken between 4640 and 4649 ft (recovery 55 per cent). The rock types include pale greyish brown calcite mudstone with vertical veins up to 18 mm wide of white and red-stained calcite, darker greyish brown fine-grained limestone, stylolitic in places, and a pale grey pellety banded calcite mudstone, the pellets possibly composed of algal mud. The larger fossils found are essentially as in the previous core with the addition of *Composita* cf. *ficoidea* (Vaughan) and *Linoproductus* cf. *corrugatohemisphericus* (Vaughan).

One foot of core was recovered from the depths 4738 to 4740 ft. It consists of pale to medium grey calcite mudstone with calcite veins and at the base a hematite-stained slickensided fine-grained limestone with *Lithostrotion?*, *Syringopora* cf. *ramulosa* Goldfuss and shell fragments.

The fourth core, 4786 to 4805 ft (recovery 16 per cent), yielded pale to medium greyish brown variable limestone including oolites, calcite mudstones and pellety organic debris limestone. Fossils determined from these beds are: *Calcisphaera sp.*, foraminifera, *Carcinophyllum* cf. *vaughani* Salée, *L. martini?*, *Davidsonina carbonaria* (McCoy) and ostracod shell fragments.

The fifth core, 4805 to 4825 ft (recovery 20 per cent), is again predominantly of a pale greyish brown fine-grained limestone, but concretionary structures, possibly of organic origin, and oolites are seen, as also are calcite veins, (up to 14 mm in thickness) and stylolites. *Calcisphaera* is abundant and other fossils include foraminifera, *Carcinophyllum vaughani*, *Diphyphyllum sp.*, *L. martini* [S$_2$ type], *Syringopora sp.*, cf. *C. ficoidea*, cf. *D. carbonaria* and *Linoproductus sp.*

Calcisphaera is the only fossil seen in the sample from the sixth core, 4868 to 4871 ft (recovery 3 per cent), which is in limestone of lithology similar to that of the previous core.

The top part of the seventh core, 4906 to 4923 ft (recovery 18 per cent), is a greyish brown calcite-veined fine-grained limestone showing several cross-sections of shells which might be *C. ficoidea*. Below this, the limestone is paler coloured, oolitic in places and rich in *Calcisphaera*. Crinoid stem columnals and Productid fragments occur sporadically.

Foraminifera and *Calcisphaera sp.* are present in the pale greyish brown calcite-veined limestone of the eighth core at 5021 to 5037 ft (recovery 12·5 per cent).

The coral *Palaeosmilia murchisoni* Milne Edwards and Haime has also been identified.

The three fragments preserved from the ninth core, 5103 to 5123 ft (recovery 10 per cent), are all of pale grey oolite with calcite veins, a stylolite, concretionary structures, and shell fragments. Foraminifera and *Calcisphaera* are present.

The three feet recovered from the tenth core, 5194 to 5209 ft (20 per cent), include deep reddish brown micaceous sandy marl at two levels, one towards the top which has numerous thin horizontal calcite veins, and the other at the base. The colour is probably due to ironstaining. The intervening limestone is calcite-veined, pale greyish brown in colour, oolitic in part and with concretionary structures, *Calcisphaera* and occasional foraminifera. Just above the lower sandy marl there is a layer of pale grey crystalline limestone.

The eleventh core, 5209 to 5225 ft (recovery 20 per cent), is composed of pale grey calcite mudstone, oolitic in part, with calcite veins and *Calcisphaera*. At the lower depth, the limestone is pale brown and finely crystalline with deep reddish brown specks, possibly of iron-stained sandy marl. A thin vertical joint filled with grey mud traverses the limestone, and a cavity in the latter contains calcite crystals and reddish brown sandy marl resembling that seen in the previous core.

The twelfth core, 5352 to 5365 ft (recovery 38 per cent), contains limestone with breccias developed at three levels. The breccias (slides E 27761–3)[1] have angular fragments, up to 4·2 cm across, of brown granular dolomite ($\omega = 1\cdot682$), of siltstone grade, some a little coarser with rhombs 0·08 mm across, turbid calcite and dolomite siltstones with fossil fragments represented by recrystallized carbonate, and green marl fragments less than 1 mm across. The matrix is a pale greyish brown fine-grained limestone. One 25·5 cm length of core is entirely composed of brecciated limestone. The interbedded limestone is a greyish brown oolite with calcite veins, and is crowded with shells identified as *Composita* cf. *ambigua* (J. Sowerby) and *C.* cf. *gregaria* (McCoy). *Calcisphaera sp.*, *Koninckopora inflata*, *Aphralysia?* and Productid fragments are also present.

The final length of core, the thirteenth, is from 5426 to 5441 ft (recovery 26 per cent). At the top is a brecciated limestone. Slides E 27764–5 cut from this breccia are composed of reddish brown granular limestone with angular fragments up to about 1·6 cm across of turbid carbonate-siltstone apparently calcite and of pale green dolomite ($\omega = 1\cdot682$) up to about 1 cm across. Shards of argillaceous rock in the main are apparently composed of kaolinite and illite. Coarsely crystalline calcite plates representing original fossil fragments and streaks of dark brown material, probably limonite, are also present. The rock is iron stained along irregular narrow zones, some possibly stylolitic. One area, of micaceous silty limestone or dolomite, contains many muscovite flakes which are outlined by limonite. Limonite has also pigmented the carbonate. Some small patches about 0·4 mm across, are composed of a coarsely crystalline carbonate centre, surrounded by a rim of pale brown granular crystals which appear to be quartz. The margin between the quartz or chert and carbonate is the hypidiomorphic outline of the carbonate crystals. Below

---

[1] Numbers prefixed by E refer to specimens in the English Sliced Rock Collection of the Institute of Geological Sciences.

the brecciated limestone is pale grey pink-stained oolite with *Calcisphaera sp.*, *Composita?* and Productid fragments.

Chippings were taken between the lowest core at 5441 ft and the base of the hole at 5588 ft. From the evidence of these, the predominantly greyish brown fine-grained limestone continues to the base of the borehole.

The fossils collected between 4630 and 4825 ft are indicative of the *Seminula* (S_2) Zone, and between 5352 and 5365 ft of the Upper *Caninia* (C_2S_1) Zone. The specimen of *P. murchisoni* from between 5021 and 5037 ft also probably indicates C_2S_1 age, the occurrence of this species in the South Western Province being C_2S_1 and D zones. The coarse breccias near the base of the hole which are associated with C_2S_1 brachiopods are probably an expression of the movement and denudation that took place in Lower Viséan times in Central England (see Hudson and Turner 1933, pp. 455–66; Hudson and Mitchell 1937, p. 12; Mitchell and Stubblefield 1941, p. 205). This appears to be the first evidence of intraformational brecciation that has been recorded from the Kent Lower Carboniferous.

The nearest locality at which Lower Carboniferous rocks have been proved is about 14 miles to the north-west at Warlingham, Surrey. In the Geological Survey Warlingham Borehole, however, all the Carboniferous Limestone (4504 ft 10 in to 5001 ft; 496 ft 2 in) is assigned to the Tournaisian and faunas characteristic of the Lower *Caninia* (γC_1), *Zaphrentis* (Z) and *Cleistopora* (K) zones have been recognized. It is of interest that the rocks in the upper part of the Carboniferous Limestone at Warlingham have been affected by dolomitization and this may provide a link with the angular dolomite fragments in the breccias of the Penshurst Borehole.   C.J.S., M.M., P.A.S.

Chapter III

CRETACEOUS: WEALDEN BEDS

GENERAL ACCOUNT

FOLLOWING ON THE Purbeck Beds, deposition of freshwater, brackish water and quasi-marine strata continued without a break, and some 2000 ft of shallow water deposits, overlying the now concealed Jurassic beds, accumulated in a gradually subsiding basin of sedimentation which extended over a considerable area of what is now the south of England and part of France and Belgium. This area is generally regarded as having been a wide lagoon bounded largely by low hills of Jurassic rocks but with a ridge of Palaeozoic rocks to the north. The deposits were deltaic, and the presence of sun cracks, fossil footprints and marsh-soil beds crowded with rhizomes of *Equisetites lyelli* indicates very shallow water conditions with emergent mudflats. The end of the Wealden episode was marked by an incursion of the sea from the south. The Wealden Series and the Upper Purbeck Beds thus form a freshwater and brackish sequence between the marine Jurassic and the marine Cretaceous rocks; no precise lithological dividing plane between the Jurassic and Cretaceous is here present (see Allen 1955, p. 266; Allen *in* Howitt 1964, p. 113; Casey 1963). The macrofossils in the Wealden Beds generally have Jurassic affinities, except towards the top of the series, where brackish water and marine forms of Cretaceous type appear. The microfossils on the other hand show the Wealden to be of Valanginian, Hauterivian and Barremian (i.e. Lower Cretaceous) age (see Anderson and Hughes 1964; Allen 1967, p. 28).

The study of the macrofossils of the Wealden Beds has been greatly neglected in this country and the following remarks by Kitchin (in Lamplugh, Kitchin and Pringle 1923, pp. 231–2) summarize the views formerly held by palaeontologists with regard to them: "The few species of molluscs usually met with are often so preserved that they do not favour the detection of any slight mutational changes that might be turned to account in stratigraphical comparisons. The freshwater forms tend to be stereotyped, while evolutionary progress and differentiation, as in the Purbeck Beds, was less rapid and of less pronounced character than in the case of marine assemblages. All these circumstances combine to retard progress in the utilization of the faunal elements in stratigraphical correlation". The progress of research, however, has brought an increasing awareness of the value of the Ostracoda found in both the Wealden and Purbeck Beds for purposes of correlation (Anderson, Bazley and Shephard-Thorn 1967, p. 189) and recent study of the bivalves of these beds has shown them to be potentially useful as indicators of horizons within broad limits.

A lithological basis of subdivision applicable to the whole Weald was adopted by Drew (1861, pp. 271–86) who named the various divisions of the series from the localities where they are best developed; these localities are all beyond the limits of the present area. Of the two major lithological divisions of the series (see Table of Strata, p. 7) the Hastings Beds consist of a group of sands and clays with subsidiary limestone and ironstone, sands predominating; they are overlain by the Weald Clay, consisting of shale and clay with subordinate beds of sandstone and limestone. The Hastings Beds have been further subdivided as follows:

$$
\text{Tunbridge Wells Sand} \begin{cases} \text{Upper Tunbridge Wells Sand} \\ \text{Grinstead Clay} \\ \text{Lower Tunbridge Wells Sand} \end{cases}
$$

Wadhurst Clay
Ashdown Beds

Around Finch Green [507415] the presence of a thick bed of sandstone, the Cuckfield Stone, divides the Grinstead Clay into an upper and lower division.

Allen (1949, p. 257) has described a thin but widespread pebble bed at the top of the Ashdown Beds, the material of which he has recently concluded came mainly from the north-east and north-west (Allen and Krumbein 1962, pp. 534–7). Further detailed work by Allen (1938–67) during the last 30 years has built up a convincing picture of the cyclic growth and subsidence of successive delta complexes in the Weald. He recognized three major cyclothems: the Ashdown Beds and Wadhurst Clay comprise the first cyclothem; the Lower Tunbridge Wells Sand and Grinstead Clay the second; and the Upper Tunbridge Wells Sand and Weald Clay the third. Where the Cuckfield Stone is present the second cyclothem can be divided into two minor ones; the lower formed by the Lower Tunbridge Wells Sand and Lower Grinstead Clay, and the upper by the Cuckfield Stone and Upper Grinstead Clay. Where the Grinstead Clay is absent in the east it is impossible to draw a line between the second and third cyclothems. Each cyclothem can be made up of eight diagnostic horizons, although locally a horizon may be absent or duplicated several times (Allen 1960). The phases of the cyclothem are thought to correspond to pulses of the Neocomian sea (Allen 1959, p. 341–2).

Another pebble bed has been found at the top of the Lower Tunbridge Wells Sand in the East Grinstead, Haywards Heath and Uckfield districts, southward of the Sevenoaks map area (Allen 1959, p. 303).

Drew separated the Tunbridge Wells Sand into three divisions from the evidence in the neighbourhood of East Grinstead in Sussex. From the remarks by Topley (1875, p. 75) it seems that Drew was not certain that the middle clay division which he termed the Grinstead Clay maintained the same horizon, and a few miles east of East Grinstead he certainly sometimes misinterpreted Grinstead Clay as Weald Clay (Buchan 1938) and sometimes as Wadhurst Clay. But Drew recorded (1861, p. 281) the occurrence of Grinstead Clay at Penshurst, and he mentioned that the clay thickens to the west.   C.R.B.

Though regional evidence shows that both Weald Clay and Hastings Beds eventually thin out northwards towards the Palaeozoic platform under London, over 800 ft of Wealden Beds were proved at Old Soar, near Plaxtol (Whitaker 1908, p. 237) and at Hurst Green a borehole which commenced in Weald Clay was on completion at 1138 ft still in clayey strata that may be of Wealden age (see p. 32).   S.C.A.H.

HASTINGS BEDS

ASHDOWN BEDS

The term Ashdown Sand was given in the first instance by Drew (1861, p. 277) to beds comprised of hard sandstones, sandrock (i.e. sand weakly cemented into a soft rock, easily broken down into its constituent grains), sands, siltstones and silts, and clays lying between the Purbeck Beds and the Wadhurst Clay; subsequently the lower part of the succession in the neighbourhood of Hastings and Fairlight, in which clays are important, was termed 'Fairlight Clays' and the term 'Ashdown Sand' restricted to the overlying sands. Recent work on Geological Sheets 303 (Tunbridge Wells), 304 (Tenterden) and 320 (Hastings) has shown that the Fairlight Clays cannot be distinguished as a stratigraphical unit, but that the lower part of the Ashdown formation contains discrete, lenticular, mappable clay beds. Accordingly the term Ashdown Beds is adopted to cover the whole of the formation. Clays for about 50 ft above the top Purbeck limestones around the Purbeck inliers of the Weald have been shown by their ostracods to contain a Purbeck fauna.

The change from the predominantly calcareous and argillaceous deposition of the Purbeck to the silty and arenaceous Ashdown Beds deposits appears to have been gradual and the two facies interdigitate so that a lithological boundary drawn between the Ashdown Beds and Purbeck Beds is frequently arbitrary (Howitt 1964, p. 81). In the present Memoir both sands and clays are grouped as 'Ashdown Beds'.

The trial bore for coal at Penshurst in 1899 (see p. 16), which started approximately 200 ft below the top of the Ashdown Beds, proved a thickness of 552 ft so that the total thickness of the formation in the Penshurst district approaches 750 ft. Both this borehole and the subsequent deeper one for oil (p. 16), show that clay, and silty shale with shelly layers and bands of clay-ironstone are common near the base of the Ashdown Beds, succeeded by shaly sandstones with some sandy shale, followed upwards again by about 200 ft of silts and sands. The highest beds are soft sandstones with a subsidiary amount of clay and shale.

Only the top 250 ft of the Ashdown Beds are exposed in the district, coming to the surface in inliers south-south-east of Edenbridge, around Penshurst and south and south-east of Tonbridge. Allen (1949, fig. 2, p. 268) has recorded the presence of the thin pebble bed at the top of the formation, noted above, around Penshurst. The localities which fall within the area of Sheet 287 are as follows: Cowden Pound; two near Eden Hall; Hever Stream south of Hever; two sides of the lane 1 mile S. of Penshurst; lane-bank north of Printstile, Bidborough; old sandpit ⅜ mile N.E. of Honnington, near Southborough; ¾ mile E.S.E. of Honnington; and at an old pit near The Grove, about 1 mile S.W. of Penshurst, in the lane leading to Salmans. From the last place Allen has described Jurassic, Carboniferous and Old Red Sandstone pebbles (1960, pp. 158, 160; 1961, pp. 279–82).   C.R.B., S.B., S.C.A.H.

WADHURST CLAY

The Wadhurst Clay comes to crop over extensive tracts in the southern quarter of the map area. The great mass comprises clay and shale, with subordinate amounts of calcareous sandstone, clay ironstone, shelly limestone, and occasional seams of lignite a few inches thick.

At the surface the clay is usually stiff and of a drab colour, brown or grey with a blue or green tinge. It is occasionally mottled grey and ochre but rarely brightly coloured except at the top where it is often brick red. 'Tilgate Stone', a fine-grained, hard, calcareous sandstone, occurs at irregular intervals throughout the formation either as lens-shaped masses or as thin continuous layers.

There are several layers of clay ironstone, the most important horizon being 20–25 ft above the base of the Wadhurst Clay, from which level it was formerly dug for iron ore for the furnaces in the Weald. Many shallow pits, from which the ironstone was removed, remain to mark the horizon and at several localities the name 'Mine-pits' still survives. Locally the 'mine-pits' (= bell-pits) can be traced for distances up to one mile. The clay ironstone is hard, smooth and light grey in colour when the iron ore is in the form of the carbonate but weathers to the oxide as a soft yellowish brown mass. Occasionally the ironstone takes the form of a matrix to a conglomerate of shell casts notably of *Neomiodon* and *Viviparus* but more frequently it occurs as small nodules or layers of nodules rarely exceeding four inches thick. Nodules are usually marked by thin concentric coatings of hydrated oxides.

Associated with the ironstone are lenses of massive unfossiliferous limestone, and thin layers of shelly limestone containing *Neomiodon* and *Viviparus*, which were dug with the ironstone and used as a flux. Similar limestones occur at other horizons. The surface of the Wadhurst Clay is pocked with old clay pits all of which are now overgrown.

Good exposures of the Wadhurst Clay are rare. A representative section which shows most of the constituent rock types occurs at the Quarry Hill brick and tile works [586450], $\frac{1}{2}$ mile S. of Tonbridge railway station, and a typical section showing the junction with Lower Tunbridge Wells Sand at the High Brooms brick and tile works [595418], about $\frac{1}{4}$ mile N. of Southborough railway station.

The average thickness of the formation within the area calculated from the surface outcrop is about 180 ft with a maximum proved thickness of 239 ft shown in a borehole at Saint's Hill.

Many of the borehole records show a thickness of Wadhurst Clay greater than that calculated from the surface outcrop. The majority of these wells are sited in the bottom of valleys and the anomalous thicknesses, 239 ft at Saint's Hill (287/50), 214 ft near Southborough (287/30) and 204 ft near Pembury (287/83), could be attributed to valley bulging [the numbers in brackets refer to the well records in the Water Supply Papers of the Institute of Geological Sciences (Well Catalogue Series) in the area of New Series One-inch (Geological) Reigate (286) and Sevenoaks (287) sheets (Cooling and others 1968)]. This compares with 195 ft near Brenchley (287/4) and 163 ft near Hever (287/12) where the wells are sited on the crests of ridges. In a well at Styles' Place, formerly Style's Place Brewery, Hadlow (Whitaker 1908, p. 150), the Wadhurst Clay is regarded as being 127 ft thick. The thickness of the Wadhurst Clay in the small inlier near Old Lodge [425415] is calculated as 100 ft. This thickens eastwards to 150 ft towards Markbeech.   S.B., C.R.B.

TUNBRIDGE WELLS SAND

In the western part of the map area it is possible to separate the Tunbridge Wells Sand into a lower and upper division by a thick bed of clay, the Grinstead

Clay. The latter attains its greatest thickness in the west and thins eastwards until east of a line running from Tonbridge to Southborough the clay is absent, except for one small outlier at Lower Green, Pembury. Over the eastern third of the map area the Upper and Lower Tunbridge Wells Sand are mapped as one horizon, Tunbridge Wells Sand (undifferentiated). For descriptive purposes this division may be included with the Lower Tunbridge Wells Sand, since many of the outcrops in the east are thin and include mainly the lateral equivalent of the Lower Tunbridge Wells Sand.

LOWER TUNBRIDGE WELLS SAND

The division maintains an average thickness of 110 ft over the western two-thirds of the map area, and crops out on either side of the Gilridge and Penshurst anticlines. The top of the division, with which the Grinstead Clay makes a clean contact, is usually marked by a massive, hard, cross-bedded sandstone, the Ardingly Sandstone (Gallois 1966, p. 47), which gives rise to many natural crags and was quarried in the past as a building stone. Typical sections are shown in Plates IIIB, IVA and B. The outcrop of the Ardingly Sandstone appears to be closely related to that of the Grinstead Clay, although similar impersistent massive beds are known from other horizons.

The lower part of the Lower Tunbridge Wells Sand bears a lithological resemblance to the Ashdown Beds and was probably deposited under similar conditions. It is composed of fine, white, yellow or brown quartzose sand and silt, locally ferruginous, and with thin beds of clay and occasional coarser seams of sand. Lignite is scarce but may be seen as fragments or as small lens-shaped spreads. Only locally are the beds calcareous.

The Top Lower Tunbridge Wells Pebble Bed has nowhere been recorded in this map area. The Grinstead Clay rests directly on the Ardingly Sandstone (Allen 1959, p. 303).

The junction with the underlying Wadhurst Clay is sharp and invariably marked by a spring line. However the boundary in the field is often obscured by downwashed sands aided by the spring water issuing from the junction.

C.R.B.

GRINSTEAD CLAY AND CUCKFIELD STONE

The outcrop of this subdivision runs from near Dormans Land in the south-west corner of the district, where it is 50 ft thick, north-eastwards to near Ockham, south-east of Marsh Green. East of this locality the bed is much broken by faulting and individual outcrops are small. From the River Eden, south of Penshurst Station, a thin, 25-ft bed of clay can be followed E.N.E. through Leigh as far as Ramhurst Manor [564467]. South of the Penshurst anticline the Grinstead Clay is well developed around Finch Green and may reach 60 to 70 ft in thickness. The thickness here, however, is largely due to the development of the Cuckfield Stone, 25 to 30 ft thick. Eastwards there is a small outcrop south of Southborough continuous with a larger outcrop on the adjacent sheet, and a small outlier, 23 ft thick, capped by the Upper Tunbridge Wells Sand at Lower Green, Pembury.

The Grinstead Clay exhibits sharp junctions with the overlying and underlying formations of the Tunbridge Wells Sand. It is characteristically composed of grey or bluish grey clay and shale, often reddened where overlain by the Upper Tunbridge Wells Sand. C.R.B.

Upper Tunbridge Wells Sand

The upper subdivision of the Tunbridge Wells Sand consists of white or yellow quartzose siltstones or fine-grained sandstone, locally ferruginous, with unconsolidated silt or sand, loam and clay. Usually the sandstone is thinly bedded or flaggy. At the base of the formation the junction with Grinstead Clay is relatively sharp. In the present map area the Weald Clay is everywhere faulted against the Upper Tunbridge Wells Sand and the full thickness of the bed cannot be calculated. It attains its greatest thickness (220 ft) around Dormans Land in the west. Eastwards, where the Upper and Lower Tunbridge Wells cannot be distinguished, the total thickness of the Tunbridge Wells Sand was 200 ft in a borehole at Styles Place (Whitaker 1908, p. 150).   C.R.B.

Weald Clay

The outcrop of the Weald Clay occupies a belt of land of low relief some $3\frac{1}{2}$ miles wide across the district, and lying 100 to 300 ft above sea level. It forms a plane sloping gently from the base of the Lower Greensand escarpment to the River Medway. Drainage from the Hythe Beds has dissected much of the slope, and locally the higher ground is drift covered. The typical appearance of Weald Clay country is shown in Plate IIB.

In the present district the full thickness of the Weald Clay is probably more than 1100 ft in the west and thins northwards and eastwards to about half that amount. A borehole [40405055] at Hurst Green made in 1929 (Buchan and others 1940, p. 31), which commenced approximately 100 ft below the top of the Weald Clay, indicates 1138 ft of clays all of which are presumed to be Weald Clay. In the Warlingham Borehole [34765719] in the area of the Reigate (286) Sheet $5\frac{1}{2}$ miles N.W. of Hurst Green a thickness of 577 ft was proved (Worssam 1958, p. 29). A boring [618540] at Old Soar, 1 mile E.N.E. of Plaxtol proved 669 ft of Weald Clay (Whitaker 1908, p. 237), and in borings near Maidstone (Worssam 1963, p. 13) approximately 600 ft were found.

The constituent materials of the formation were largely deposited in deeper water than were the sandy members of the Wealden Series, but the beds show locally a persistence of shallow water conditions. They consist chiefly of thinly bedded shales, mudstones and silty mudstone all of which weather to a predominantly stiff clay. These clays and shales are usually pale or dark grey tinged with blue or brown, and near the surface they weather through a mottled zone to brown or yellow. Red clays occur but are rare. At intervals within the clays are beds of shelly limestone, and of sand and sandstone. Occasionally clay ironstone occurs as nodules. These subordinate beds are rarely more than a few inches thick and as a rule cannot be traced individually for any great distance. Reeves (1968, p. 466) was unable to recognize the red clays of his Group II within the present map area. He thought that all three groups thinned in a west to east direction. However, it is possible that Group I, 330 ft thick at Horley, is locally cut out by the major fault which in the district forms the southern limit of the Weald Clay (see p. 10). The exact throw of this fault is unknown, but it may be of sufficient magnitude to account, in part, for the absence of the red beds in Group II.

Ostracods are distributed generally throughout the beds, often very abundantly and form the only basis for its subdivision into faunal zones (Anderson *in* Shephard-Thorn and others, 1966, pp. 82–9; Thurrell, Worssam and Edmonds

B. ARDINGLY SANDSTONE, NEAR CHIDDINGSTONE HOATH

(A 6812)

A. BASAL BEDS OF THE ARDINGLY SANDSTONE, NEAR SPELDHURST

1968; Anderson 1967); mollusca are of more sporadic occurrence and are found chiefly in the limestones, many of which are crowded with the shells of the gastropod *Viviparus* [*Paludina*] or the bivalve *Filosina* [*Cyrena*].

There are three main types of limestones, namely Small-'*Paludina*' limestone, composed largely of shells of the small gastropod *Viviparus elongatus* (J. de C. Sowerby), Large-'*Paludina*' limestone formed from the larger species *V. sussexensis* (Mantell) and '*Cyrena*' limestone with the brackish water bivalve *Filosina gregaria* Casey as its chief constituent.

Topley (1875, p. 102) considered that in the Weald Clay there were seven horizons of either sandstone or limestone, although not all of them are known in the present map area. Recent work has shown that some modifications in detail to his scheme are necessary. Topley's beds in ascending order are as follows: 1, Horsham Stone; 2, sand and sandstone; 3, Small-'*Paludina*' limestone; 4, Large-'*Paludina*' limestone; 5, sand and sandstone; 6, Large-'*Paludina*' limestone; 7, sand.

Bed 1, the Horsham Stone, does not persist east of Crawley on the Horsham (302) Sheet.

The existence of Bed 2, a sand or sandstone, is doubtful. There is no evidence for the bed in either the western part of the Weald (Worssam and Thurrell 1967, p. 264; Thurrell, Worssam and Edmonds 1968) or the eastern end (Smart, Bisson and Worssam 1966, pp. 47–8). A sand at Leigh, 2 miles W. of Tonbridge, suggested by Topley (1875, p. 104) to belong to Bed 2 is in fact part of the Upper Tunbridge Wells Sand.

Bed 3, the Small-'*Paludina*' limestone is present in the area of the Sevenoaks Sheet. Small-'*Paludina*' limestone was found near Bowerland Farm [395459] 1¼ miles N. of Lingfield Station and is probably on the same horizon as that mapped by Dines on the Reigate (286) Sheet to the south of Crowhurst. Other localities where Small-'*Paludina*' limestone has been recorded are: near Bough-Beech Place (formerly Ivy House [491475], but now demolished); near Kilnhouse Farm [501482]; near Hall's Green [527494]; in Starvecrow Wood [597495] 2 miles N. of Tonbridge (see p. 51) and at East Peckham some 5 miles E.N.E. of Tonbridge. It was from this last locality that one of the two syntypes of *Viviparus elongatus* figured by J. de C. Sowerby, a slab of *Viviparus* limestone, was obtained (Arkell 1941, p. 116).

Bed 4, the lower of the two horizons of Large-'*Paludina*' limestone described by Topley, is only present on the south side of the Weald.

Bed 5 is probably present in the Sevenoaks district though only one exposure, 20 ft of sand and sandstone, in the railway cutting [42234867] west of Batchelor's Farm (Topley 1875, p. 106) was definitely assigned to it.

Bed 6, the upper horizon of Large-'*Paludina*' limestone, has been mapped for a short distance around Sevenoaks Weald and around Goldings 2 miles E. It has been recorded in the Sevenoaks and Tonbridge Railway cutting where it was 22 in thick and, in places, in two beds; on Hadlow Common due west of Goose Green; and at Budd's Green, 1 mile W. of Shipbourne, formerly Shipborne, (Topley 1875). One other locality where Large-'*Paludina*' limestone has been recorded, but at a lower horizon (? Topley's Bed 4), is in the stream [52854882] 550 yd W. of Southwood. It here occurs 650 yd downstream from, and presumably stratigraphically below, an outcrop of Small-'*Paludina*' limestone [52754934]. This relationship could possibly be accounted for by folding or faulting, but in the area of the Reigate (286) Sheet Small-'*Paludina*' limestone has been recorded in the Newdigate Brickworks high in the succession (Worssam and Thurrell 1967, p. 265).

Bed 7, a sand, has not been recognized in the Sevenoaks district.

In general, on present data, the precise stratigraphical horizon of the considerable number of exposures of silts, sands, sandstone and limestone additional to those known to Topley, recorded below, cannot in fact be determined.

The highest part of the Weald Clay is of near-marine facies comparable with that of the 'Cinder Beds' of the Purbeck, and contains foraminifera, echinoid spines and the molluscs *Cassiope, Ostrea, Corbula, Nemocardium* and *Filosina* (Casey 1961, p. 490). Within the present map area the only record of the higher strata is that by Caleb Evans (1871, pp. 1–3) from the Sevenoaks railway tunnel. His faunal list is brief but includes *Ostrea, Cardium* and *Cerithium* or *Potamides*. This gastropod is possibly *Cassiope*. The brackish water *Filosina* is common in the beds below. Chatwin (*in* Dines and Edmunds 1933, p. 115) gave a more detailed faunal list from the highest Weald Clay beds exposed in the Redhill–Earlswood railway cutting to the west of the present map area.

Cassiope is a marine to brackish water gastropod and probably lived under conditions of salinity similar to those which suited the brackish water *Filosina* (Casey 1955) which makes the beds of '*Cyrena*' limestone. *Cassiope strombiformis* Schlotheim has been found at Starvecrow (see p. 51) at the horizon of the Small-'*Paludina*' limestone. *Cassiope* identified as "*Glauconia* cf. *lujani*", at possibly the same horizon, has been recorded as occurring with '*Cyrena*' and Small-'*Paludina*' limestone just west of the Sevenoaks district in a brickyard [393464] at Crowhurst, on the eastern margin of the area included in the Reigate (286) Sheet (Dines and Edmunds 1933, pp. 38–9). In the area north of Horsham *Cassiope* bands are known towards the top of Topley's Bed 3, and at the Newdigate Brickworks, towards the top of Topley's Bed 5. These appear to link with two brackish-marine bands in the Warlingham Borehole (Worssam and Thurrell 1967, p. 264).

The base of the Weald Clay is not seen, as everywhere within the map area the Weald Clay is downfaulted against the Hastings Beds or hidden beneath superficial deposits.   C.R.B., C.J.W., S.B.

PETROGRAPHY OF THE WEALDEN BEDS

The arenaceous members of the Hastings Beds, i.e. the Ashdown Sands and the Lower and Upper Tunbridge Wells Sands comprise fine sands, sandy silts and silts; sandy beds also occur locally in the Grinstead Clay. Somewhat coarser material occurs in the Lower Tunbridge Wells Sand than in the other divisions, and an indication of the range of average grain-sizes, based on micrometric measurements, is given in Table 1. Quartz, in subangular grains, is everywhere the dominant constituent, seldom making less than 95 per cent of the sand or silt. Although generally in an incoherent condition, in places sufficient cementing material (mainly limonite) is present to give rise to sandrock. The colour varies, according to the amount of limonite present, from white to buff or pale brown. Minor constituents include grains of quartzite, chert and

TABLE 1
Grain Counts on Heavy Residues from the Wealden Sands

Sample No.	Horizon	Grain size (mm)	Magnetite ilmenite	Pyrite	Leucoxene	Limonite	Zircon	Tourmaline	Rutile	Glauconite
ASHDOWN BEDS										
3	High	0·02	15	18	54	5	7	X	X	—
16	High	0·06	3	X	72	X	9	9	5	—
17	High	0·08	8	2	35	X	28	10	16	X
23	High	0·07	4	X	42	X	20	X	6	28
29	High	0·02	7	—	49	X	15	18	9	—
LOWER TUNBRIDGE WELLS SAND										
24	Ardingly Sandstone	0·07	7	X	44	X	18	3	6	21
1	,,	0·20	22	—	64	3	6	5	X	X
6	,,	0·30	16	X	60	10	10	3	X	—
15	,,	0·25	10	6	58	6	9	4	7	X
18	,,	0·02	4	—	38	X	15	19	20	—
19	,,	0·05	9	X	32	12	38	3	6	X
21	,,	0·20	13	5	58	7	2	8	7	—
22	,,	0·20	11	2	52	X	24	3	8	X
26	,,	0·10	5	X	59	X	20	9	4	2
28	,,	0·30	10	—	33	X	42	8	7	X
2	,,	0·20	14	—	43	—	26	8	8	—
25	Low	0·10	6	X	50	9	10	19	6	X
30	Low	0·03	5	X	60	X	15	14	6	—
4	Low	0·15	17	3	57	X	13	4	3	—
5	Low	0·05	7	X	66	X	7	15	5	—
8	Middle	0·10	6	—	79	7	7	X	X	—
9	Middle	0·07	X	X	62	X	19	9	X	9
13	Low	0·04	2	—	52	X	12	17	13	—
14	Low	0·10	8	X	68	8	10	6	X	X
31	Low	0·20	18	X	68	X	9	2	3	X
35	Middle	0·06	9	—	36	X	38	9	8	—
36	Middle	0·03	8	—	26	—	54	2	10	X
32	Tunbridge Wells Sand, undivided	0·05	10	—	59	—	6	22	2	—
33	,,	0·10	11	—	64	—	12	8	4	X
CUCKFIELD STONE										
20		0·06	X	8	55	X	25	2	10	X
UPPER TUNBRIDGE WELLS SAND										
7	Middle	0·01	5	—	35	X	22	19	16	—
10	Low	0·04	2	—	44	9	16	13	10	—
11	Middle	0·01	5	—	32	X	17	17	25	—
12	Low	0·01	4	—	25	X	20	23	20	—
27	Middle	0·08	4	—	51	X	30	10	4	—
34	Low	0·20	2	X	74	—	7	8	6	X
SAND BED IN WEALD CLAY										
37		0·05	19	7	47	X	18	4	4	X

X = less than 2%
— = not observed

D

mudstone, occasional flakes of muscovite, fine micaceous aggregate and, very rarely, feldspars. Minerals with specific gravity greater than 2·76 are present in amounts varying between 0·002 per cent and 0·2 per cent by weight. Separations of these were made from 37 samples (E 21612–21648) with the object of assessing the possibility of distinguishing between the three arenaceous divisions, and the results of counting a total of over 12 000 grains (an average of approximately 350 per sample) in the residues obtained are given in the accompanying tables.

The heavy mineral suite is dominated by leucoxene, with zircon, tourmaline, magnetite, ilmenite and rutile as persistent constituents. In general these minerals show a more worn appearance than the quartz grains which make up the bulk of the sands. Both prismatic and zoned zircons are rare, though not entirely absent. Well-worn grains of purple zircon were noted in small numbers in most of the residues, and there were a few yellow grains of this mineral. The presence of monazite was not confirmed. Prismatic tourmaline is rare except in the finest grade of silt; shades of brown and green predominate, with a few blue and bluish grey types. Both yellow and red varieties of rutile occur, numerous examples of worn geniculate twins being observed. Glauconite is sporadically present in fresh or partly oxidized pellets. Pyrite, evidently authigenic from its ragged form, is also sporadically distributed. Anatase, both yellow and bluish, occurs in well-formed crystals. Brookite was noted in 13 samples. Minerals recorded as rare or isolated grains include kyanite, garnet and staurolite (all three divisions), hornblende, sphene and chlorite (Lower and Upper Tunbridge Wells Sand) and corundum (Lower Tunbridge Wells Sand). "Shimmer aggregate" micaceous material which has possibly replaced cordierite or staurolite, was recorded in a few instances. The whole suite confirms the view expressed by Milner (1923, p. 296) and confirmed by Allen (1949, pp. 304–5; 1954, pp. 500–7) that the Hastings Beds were derived from pre-existing sediments, and not directly from crystalline rocks; but neither the grain counts nor a study of varietal types among the minerals provided any satisfactory basis for discriminating between the three arenaceous divisions in the present area. One sand from the Weald Clay (E 21648) proved to have a similar heavy mineral suite.

K.C.D.

DETAILS

Ashdown Beds

In the inlier between Crippenden [447418] and Ludwells [455417] 2¾ miles S. of Edenbridge, on the southern flank of the Gilridge anticline, sections revealed by deeply entrenched streams show that the Ashdown Beds there are thin flaggy silts and sandstones. These thin beds are relatively incompetent and have been affected by valley bulging in the stream to the north-west of Ludwells.

Nearer Edenbridge, in another inlier between Cobhambury [451431] and the railway line, fine-grained flaggy sandstone and siltstones are exposed in the stream (Warren Gill) on the eastern side of the Edenbridge road and in the railway cutting. Warren Gill is floored by sandstone for most of its length between [46494341] and the railway line. At 200 yd S.W. of the railway line [468440] a six-foot high waterfall is formed by the Ashdown Beds sandstones.

Thin flaggy sandstone dipping 5° N.N.E., was exposed in the old pit [46744378] south-east of the Warren.

The railway cutting, together with the stream running along the eastern side of the railway, provides a continuous section of silts, sands, thin flaggy siltstones and sandstones between the northern end of the tunnel [473432] and the small stream which crosses the line at [471439].

An old pit [45204303] south-east of Cobhambury shows the uppermost beds of the Ashdown Beds to be massive sandstones which have weathered flaggy. The dip of these beds is 12° N.W. Up to 6 ft of yellow flaggy siltstone dipping 7° N. 80° E. were noted by Dr. Buchan in 1933 in the old pit [45704327] 600 yd N. 80° E. of Cobhambury. In 1965 this pit was completely overgrown.

Another inlier between Falconhurst [46904255] and Cowden Station has Ashdown Beds well exposed in the railway cutting and in the stream south of the station. Thin flaggy sandstone was vertical at 47644166, while thicker sandstone 100 yd S.E. was horizontal. At 200 yd farther downstream the flaggy sandstone dips 45° S.W. One other dip in the stream section was of 70° N.E. 350 yd N.W. of the station [47444195]. All these anomalous dips are attributed to valley bulging. The dip of 3° S.E. recorded on the eastern side of the railway cutting is in accord with the general 1½° S.E. dip for this inlier.

The fringe of the large inlier around Penshurst contains pale greyish yellow flaggy sandstone with a small amount of ironstone. The following section, in downward sequence, was noted in a small pit [55474395] on the west side of the road 500 yd N.E. of Printstile: sand and sandstone rubble, 3 ft; pale grey sandstone (in beds up to 4½ in thick, ferruginous along open bedding planes and joints), 3 ft; buff clayey sandstone (in beds up to 2 in thick) locally ferruginous, soft and powdery when dry, 3¾ ft; clay, 3 in; massive sandstone, 2 ft.

Allen (1960, p. 158) recorded the Top Ashdown Pebble Bed from an old quarry [513429] 400 yd S.S.E. of Salmans.

Near the top of the Ashdown Beds, in Ashour Wood [54554370], 600 yd W.N.W. of Printstile occurs the following sequence: sandstone rubble mixed with downwash clay, 3 ft; pale yellow and grey, ripple-marked, fine-grained, soft sandstone in 9-in blocks, 3 ft; closely jointed sandstone with ferruginous fillings, 2 ft; hard sandstone, 4 ft.

On the northern side of the anticline a narrow outcrop of Ashdown Beds is found between the Wadhurst Clay and the First Terrace of the R. Eden 400 yd S.E. of Chantler's Farm. A disused quarry on this outcrop [51074476] exposes 3–4 ft of thin flaggy sandstone dipping 3° N.N.E. An old overgrown quarry at Wat Stock [509439] still shows a few feet of thin, flaggy, ironstained sandstone.

A small inlier of Ashdown Beds is exposed in the bed of a stream [54704305] 800 yd S.W. of Printstile. In the eastern branch of the stream thick flaggy sandstone was noted.

Sandrock is exposed in old pits [56454450] 600 yd S.E. of the Water Works at Upper Haysden.

In a further inlier of Ashdown Beds south-south-east of Tonbridge the beds have been worked at several localities but in most of the pits the sections are now obscured. At its northern edge, and at the southern end of the railway tunnel, the junction of the Ashdown Beds with Wadhurst Clay is poorly exposed. In the cutting 200 to 300 yd S. of the tunnel the beds are gently folded and separated on the south from Wadhurst Clay by a normal fault. The best local section was noted in the railway cutting, between the end of the tunnel and this fault. Prestwich and Morris (1846, pp. 398–400) saw and described this soon after it was exposed. Their record, epitomized and given its geological classification is as follows:

Ft

Wadhurst Clay

Dark-coloured laminated clays and shales   ..   —

Three beds of light-coloured sandstone separated by dark brown clays and containing *Unio gualterii* J. de C. Sowerby, *Bairdestheria subquadrata* (J. de C. Sow.) and imperfect impressions of plants. (Conformable in general stratification with the beds below but resting on a slightly uneven and waterworn plane)   20

White clayey sand (thickness not recorded)   ..   —

Black lignite clay   2

Ashdown Beds

Thickly bedded light coloured sandstones, sometimes massive, at other times fissile, and occasionally exhibiting false stratification   31–41

Sandrock, slightly ferruginous, siliceous and compact   10

Lignite clay ..   .. 8 in to 1

Sandrock, slightly ferruginous, siliceous and compact   10

Dark soft shaly sandstones   8

Sandrock   12

They recorded the dip of the argillaceous beds (i.e. Wadhurst Clay) as 2° or 3° N. They also described the highest beds of the Ashdown Beds at the southern end of the same inlier, as exposed in a railway cutting south of Tonbridge, ¾ mile S. of the above fault. Slightly annotated, with thicknesses added and classified geologically, the sequence runs:

Ft

Wadhurst Clay

Dark grey laminated clays full of *Neomiodon* [*Cyrena*] *medius* passing downwards into a lightish green clay, and then into greenish grey shaly clays full of *Cypridea spp.*, and with numerous bands of concretionary and nodular calcareo-argillaceous ironstones full of *Unio gualterii* J. de C. Sowerby, *Neomiodon medius*, *Viviparus* [*Paludina*] *elongatus* (J. de C. Sow.), and ostracods; also thin slabs of sandstone covered with vermiform impressions   30

Ashdown Beds

Whitish soft massive sandstones, with *Neomiodon* [*Cyrena*] and vegetable impressions common at the top ..   ..   ⎫

Soft white decomposing argillaceous sandstone   ..   ⎬ 12

White massive sandstone, with probable traces of grit ⎭

Dirty white soft fissile argillaceous sandstone; lower part more massive; stratification very irregular and slightly unconformable to the overlying bed   about 15

The dip is given as about 2° S.

East of the railway and about 120 yd W. of Forest Farm [59524392] the following section, in downward sequence, of Ashdown Beds, dipping 8° S.E. was noted in 1936: sandy clay mixture with fragments of fine-grained sandstone and lenses of thin flaggy sandstone, 3 ft; pale brown thin flaggy sandstone, 2½ ft; brown thin flaggy sandstone with ironstone nodules, 9 in; grey thin flaggy fine-grained sandstone, 1½ ft; pale grey massive cross-bedded sandrock with irony partings, 4 ft; white thinly bedded shattered fine-grained sandstone, 1 ft; pale grey massive cross-bedded sandrock with yellow ferruginous partings, seen to 4 ft. Elsewhere in the inlier massive sandstone predominates, interbedded with thin flaggy or fissile soft sandrock and associated with sand and loam; a typical section can be seen in the old pit [58874448] 250 yd E.S.E. of Mabledon Farm.

South of Tudeley there is one small, faulted inlier indicated by a section in a partly overgrown and infilled quarry [618450] 250 yd E. of Park Farm. This section showed 3 ft of thinly bedded sandstone, on 3 ft of thin, irregularly cross-bedded, micaceous sandstone, strongly jointed, associated with subordinate calcareous mudstone, followed by two beds, each 1 ft thick, of buff massive fine-grained sandstone.

These beds dip 10° N.E. They were noted again, in an exposure in the stream bed [62054493] 300 yd E.S.E. of the above, where the uppermost beds are thick massive flat-lying sandstones flooring the bed of the stream. Wadhurst Clay was exposed in the ditches in the fields immediately to the east of the stream. Thinner, flaggy sandstone occurs farther upstream at 61934460.

Yet another small inlier lies a little over ½ mile N. of Brenchley church and is exposed in the stream [681427] and road cutting [68054275], where the most northerly and highest beds are yellow fine-grained sandstones showing ripple marks and with a dip of 15° N. Beneath these is a predominantly silty layer with some sandstone. The lowest beds seen were massive fine-grained hard sandstones with cuboidal jointing. Southwards the dip decreases, being only 5° N. in the east side of the cutting on the main road, where the beds are seen to underlie the clay.   C.R.B., S.B.

WADHURST CLAY

South-west of Crippenden fragments of silty shale, ironstone and calcareous grit occur in weathered clay but the contributory beds were not seen *in situ*. Fragments of ironstone with *Neomiodon medius* and smaller species of *Neomiodon* were noted on the edges of several overgrown pits between Claydene and Gilridge, and again west of Claydene, and limestone was seen in a pit [46034151] 400 yd S.W. of Claydene, west of which dark grey shale lies within a few inches of the surface.

Bell-pits can be traced through the wood (Minepit, 445417) to the south of Crippenden Manor, separated by a small N.E.–S.W. fault from a similar line of pits in the wood (Liveroxhill, 447413) which lies to the south of the present map area on the Tunbridge Wells (303) Sheet. Bell-pits can also be found in the wood (Cobhambury, 452427) 450 yd S. of Cobhambury. In all these localities the pits lie 20–25 ft above the base of the Wadhurst Clay.

Valley bulging brings up the Wadhurst Clay from beneath the Lower Tunbridge Wells Sand in the stream [45654375] 200 yd S.S.E. of Brook Street. At 250 yd upstream nodular ironstone is associated with grey shale.

Yellow, buff and grey mottled clay overlies dark grey shale east of Cowden railway station; some 600 yd N.N.E. it contains shelly limestone. Near Hever, shelly limestone associated with stiff grey clay weathering to a buff colour was seen in the stream bed [49024407] 300 yd S. of Lock Skinners and in an old pit [50124470] 300 yd E. of Hill Hoath.

Thin siltstone and limestone interbedded with grey shaly clay cropped out in the bed of the stream (Bilton's Gill, 48604214) 800 yd E.S.E. of Horseshoe Green.

Around the Penshurst anticline the Wadhurst Clay appears to maintain a constant thickness of about 180 ft. The only deviations from this are to the south of Harden [51104295], where the thinning of the Wadhurst Clay can be attributed to strike faulting, and in the well record at Saint's Hill (see p. 30). Bell-pits, 20–25 ft above the base of the Wadhurst Clay were noted in the wood (Russell's, 507433) 550 yd W. of Salmans and in Ashour Wood at two localities [541433 and 54554370].

Wadhurst Clay is exposed in the bed and cliffs of the River Medway to the east of Smart's Hill. At 500 yd E.S.E. of Nashes [53254184] 6 ft of alluvium overlie 2 ft of gravel which in turn rest on the Wadhurst Clay; 100 yd downstream [53294192] 10 ft of alluvium rest on grey Wadhurst Clay. Mottled grey and yellowish grey shaly clay floors the bed of the Medway by the footbridge [53354206] 600 yd E. of Nashes and 60 yd downstream 5 ft of alluvium were noted overlying the Wadhurst Clay.

Thin flaggy siltstone on grey shaly clay was exposed in the degraded face of the old pit [50604522] 550 yd E.N.E. of Chiddingstone church in the small faulted inlier north of the village.

West of Bidborough stiff dark clay weathering to yellow and buff overlies shale. Thin limestone beds were noted together with siltstone but not seen *in situ*, 370 yd S.W. of Petersbank, formerly Hodges Bank [54184257], and shelly limestone in stream sections 400 yd N.E. of Petersbank. Thin beds of sandstone and shelly limestone occur in roadside pits at the road junction just east of Upper Haysden [565448]; bluish grey unfossiliferous limestone in an old pit some 400 yd N.E. of Nightingale Farm [58234425]; while numerous thin slabs of sandstone, not seen *in situ*, were noted in another old pit between the latter and the main Tonbridge–Southborough road.

Trial boreholes for the Tonbridge By-pass have proved that Wadhurst Clay underlies the Alluvium and Brickearth of the River Medway at a depth of 19–38 ft (see p. 138), between Haysden and the present river bed 800 yd N.W.

The following description of an extensive section exposed at one time and recorded by Lewis Abbott (1907b, p. 98) at the Quarry Hill brick and tile works, Tonbridge (see p. 44) has been emended slightly: yellowish clay, with ostracods, 2 ft; green shaly clay, 2 ft; limonitic shaly clay, with ostracods, 6 in; green shaly clay, 4 ft; shaly clay, 2 ft; green shale, 1 ft; on green and yellow shales with layers of ironstone 2 to 4 in thick and 4 layers of shells 1 to 2 in thick, 12 ft; hematitic layer with *Neomiodon*, 3 in; brown and yellowish brown shales, 5 ft; clay ironstone nodule bed, 6 in; on black carbonaceous clay, breaking with a short hackly fracture, with many Ostracoda, *Equisetites lyelli* and bands of limestone with *Viviparus* (chiefly *V. sussexensis*) and *Neomiodon sp.*, and a layer of cone-in-cone, $\frac{3}{4}$ in thick, seen 20 ft; "similar beds", pierced in well, 100 ft.

Bluish grey and brown clay and shale with inconstant bands of '*Cyrena*' and '*Paludina*' limestone and of clay ironstone are still to be seen. The well referred to above is in fact 125 ft deep and is reported to have passed through approximately 65 ft of Wadhurst Clay before entering Ashdown Beds. The dip of the clay in the northern part of the pit is 5° N., but at the southern end of the pit near the fault, the dip changes to 6° S. and the clay is faulted against Tunbridge Wells Sand.

Red and grey clay, with a 1-in thick bed of ironstone was exposed in the pit in the wood (Priory Wood, 591450), 450 yd W. of the Vauxhall Inn. The ironstone is probably the iron-bed 20 to 25 ft above the base of the Wadhurst Clay, which was formerly worked in the wood (Minepit Wood, 582441), 200 yd N.E. of Nightingale Farm and in the woods south of Forest Farm [596439] (see p. 41). In the northwestern corner of this pit up to 6 ft of reddish brown sandy wash overlay the Wadhurst Clay. The pit is used as a rubbish tip and the section obscured.

The upper beds of the Wadhurst Clay exposed in 1846 at the northern end of the Tonbridge railway tunnel were described by Prestwich and Morris (1846, p. 398), who noted 30 ft of brownish laminated clay overlying 20 ft of dark coloured laminated clays and shales with thin, tabular or ferruginous and concretionary, dark impure argillaceous limestone, and some thin tablets of sandstone. The authors recorded '*Cyrena*' = *Neomiodon medius*, *Viviparus elongatus* and *Cypridea spp*. in the limestone and also in the clays.

Dark buff clay containing limestone with *Neomiodon* and small *Viviparus* occurs in an old pit [59504505] between the Vauxhall Inn and the southern end of the Tonbridge railway tunnel.

At the Castle Hill brick works [605443] now disused, 900 yd S.E. of Bournemill Farm, a lens of grey, massive unfossiliferous limestone, 3 ft thick, was exposed in fine-grained yellow sandstone which occurred as thin, rather fissile, even beds. In the floor of the pit there were numerous fragments of '*Paludina*' and '*Cyrena*' limestone but neither was seen *in situ*. Ripple-marked sandstone has been recorded from this pit (Handcock 1914, p. 57).

The section of the basal beds of the Wadhurst Clay formerly seen in the railway cutting 700 yd S.W. of Forest Farm has already been described (see p. 38). A line of bell-pits, 20 to 25 ft above the base of the Wadhurst Clay can be traced through the woods (Burnt House Shaw, 599438; Grove Wood, 595437 and Minepit Wood, 593432) to the south and east of Forest Farm. Old Forge Farm [595429], as its name implies, is listed by Straker (1931) as the site of one of two forges in this vicinity. The forge pond, now silted up, was later used for a powder mill, and then a corn mill.

About a mile farther south the railway cuts the higher beds of the Wadhurst Clay and some of the overlying Lower Tunbridge Wells Sand. The section, no longer exposed, was described by Prestwich and Morris (1846, p. 401) as showing 20 ft of soft, marly, red and ferruginous sandstone, with subordinate beds of whitish sand and red clay (Lower Tunbridge Wells Sand) overlying 15 ft of Wadhurst Clay (see p. 45).

The High Brooms brick pit [594418], ¼ mile N. of Southborough, exposed the lowest 50 ft of the Lower Tunbridge Wells Sand and the upper 65 ft of the Wadhurst Clay. The junction of these two formations is marked by a colour change in the clay. The normal bluish grey colour passes up into a greenish grey clay up to 2 ft thick. Immediately underneath the sands the top 6 in of the clay are brick red in colour, although locally this colour may be absent and the sands rest directly on greenish grey clay. The marked spring issuing from the junction causes a serious water problem in the pit. Fossils including ostracods, mollusca and a vertebra of a Saurian were recorded from this pit by Handcock (1910, p. 521). This section is the type exposure for the High Brooms *Equisetites lyelli* Soil-Bed 30 ft down in the Wadhurst Clay, which was also recorded 3½ miles E. (Lock 1953, p. 31).

Slabs of blue shelly limestone were abundant in the stream east of Old Forge [600427]. The cutting for the diverted stream [60254254] immediately west of the sewerage farm exposed up to 10 ft of grey shaly clay with thin (¼ to ½ in) limestone bands.

In the stream bed (Devils Gill, 618442), 700 yd W. of Knowles Bank, '*Cyrena*' limestone occurs in association with clay and shale. Ditch sections at 62084490 in the eastern side of the stream revealed large slabs, 3 to 4 in thick, of '*Cyrena*' limestone.

'*Paludina*' limestone and sandstone were also found with clay and disturbed shale in a stream section [62884330] 360 yd N.E. of Kent College, formerly Hawkwell Place, and calcareous sandstone was noted 200 yd S.E. of Bouncers Bank [63334388].

'Tilgate' Stone on blue clay was exposed in the stream bed [64674242] 400 yd N.E. of Albans, and, within the same inlier of the Wadhurst Clay '*Cyrena*' limestone was seen 550 yd S.W. of Crittenden [65324320]. S.B., C.R.B.

LOWER TUNBRIDGE WELLS SAND

(a) Northern limb of the Penshurst–Gilridge Anticline.

In the neighbourhood of Dry Hill Camp [433418], in the western part of the map area, the Lower Tunbridge Wells Sand is 120 ft thick. The Ardingly Sandstone, about 40 ft thick, is not well developed and gives rise only to a low feature. The outcrop can be followed on to the adjoining Tunbridge Wells (303) Sheet where the sandstone forms low crags.

Deep ditch sections [43204203] to the east of Dry Hill House showed a succession of silts and thin sandstones, while flaggy sandstone, locally massive, was noted in the ditch to the north-east of the house [43104235]. An old pit [43584222] 750 yd N.E. of the Trigonometrical Station on Dry Hill showed 6 ft of sandy wash, overlying 6 ft of flaggy sandstone which in turn overlie 12 ft of massive sandstone. Thin flaggy sandstone dips 25° N.W. in the old pit [43814234] 300 yd S. of Greybury. The dip is due to faulting (see p. 10). A small anticline associated with the same fault (see p. 10) affects the thin flaggy sandstone in the stream [44164306] 100 yd S. of Clatfields, Some 6 ft of silts and siltstone associated with 6 to 12 bands of sandstone are exposed in the pit [44404336] 150 yd W.N.W. of Ockhams.

Nearer Hever, at the northern end of an old quarry [44524360] 300 yd N. of Ockhams were seen 4 ft of flaggy, fine-grained sandstone, weathering to loamy clay but still showing traces of bedding, underlain by 1 ft of pale grey, fine-grained sand, then 2 ft of fine-grained ferruginous sandstone closely jointed into small cubes, this resting on 9 ft of massive, cross-bedded, fine-grained sandstone.

Another old quarry [459439], about 200 yd E. of Brook Street, showed: clayey silt, 1 ft; yellow massive ferruginous sandstone, 2 ft; hard flaggy ferruginous sandstone, 1½ ft; white soft massive quartzose sandstone, 10 ft; yellow fine-grained flaggy ferruginous sandstone, 1 ft 10 in; yellow fine-grained hard massive sandstone, 7 in; yellow fine-grained hard massive ferruginous sandstone, 2½ ft.

Between Brook Street and Bower Farm [471449] the outcrop is much faulted. A few inches of thin flaggy sandstone can be still observed in the overgrown pit [46254392] 500 yd E. of Brook Street and thick sandstones, dipping 30° N. in the stream [45864405] 200 yd N.E. of Brook Street. Massive sandstone is still visible in the roadside pit [46804445] 300 yd E.S.E. of Hever Station.

Some 250 yd S. of Hever Castle [47774498] 18 ft of yellowish brown flaggy sandstone showing ripple marks were noted. The dip was 45° N.W. in this and the two pits to the east [47944503 and 48024506]; the steep north-westerly dip is caused by faulting (see p. 10). Immediately south of the first pit red clay was dug in a shallow pit and 350 yd S.S.E. of the Castle [47954495] soft fine-grained sand with a uniform grain 0·05 to 0·11 mm was dug. About 20 ft of sand rock, identical with the Ardingly Sandstone, are visible in the old quarry [48824505] 700 yd N. of Lock Skinners but this cannot be traced through the wood (Park Wood) and as it appears to be within 20 to 25 ft of the base of the Tunbridge Wells Sand it is at too low a horizon to be the Ardingly Sandstone.

Topley (1875, p. 89) described a section at Chiddingstone as follows: "In a quarry half-a-mile west of Chiddingstone church there is some decomposed calcareous sandstone, one layer of which contains a great number of fossils: *Unio* (large species), *Cyrena*, *Paludina*, *Cypridea* and fish. The calcareous bed has been about two feet thick, but is undecomposed for only one or two inches in an irregular line; the rest is now a sandstone which sometimes exhibits concentric wavy lines of infiltration or decomposition". The quarry [49404525] was considerably overgrown in 1937, when massive and flaggy sandstones were to be seen, but not the fossiliferous calcareous bed.

Numerous small exposures around Chiddingstone showed for the most part loam, unconsolidated sand or thinly bedded sandstone. Ardingly Sandstone was noted on the eastern side of a lake [49974517] 150 yd S.W. of St. Mary's Church; behind the

Castle Inn (the Chiding Stone, Plate IVB), and in a pit [50504507] 500 yd E.S.E. of the church where the rock was softer and clean except for irony partings at intervals. Some 70 per cent of the grains making up this last sandrock are between 0·1 and 0·15 mm in diameter; the largest grains being 0·55 mm.

An outlier of Ardingly Sandstone [495445] south-west of Hill Hoath is well exposed in the old quarry [496446] 300 yd W. of the house and as a line of natural crags [494444] 600 yd S.W. of the house.

Drew (1861, p. 280) referred to "a fine line of rocks at Redleaf, Penshurst" but considered them to belong to the Upper Tunbridge Wells Sand; they are in fact Ardingly Sandstone crags. Up to 24 ft of the massive cross-bedded sandstone form a ridge parallel to the base of the Grinstead Clay some 300 yd N.N.W. of Redleaf House (now demolished). About the same distance to the south-east of Redleaf House another line of crags [52604515] showed the following downward section: loamy clay, sand and sandstone rubble, 2 ft; loamy decomposed sandstone, 3 ft; lignite, 2 to 6 in; decomposed sandstone with indistinct bedding, 2 ft; massive sandstone, 1 ft; decomposed loamy sandstone, 1½ ft; massive sandstone, harsh to the touch, at the top of which there is an impersistent layer of lignite, 2 ft. Some 60 yd S. of this section the Ardingly Sandstone is faulted against Wadhurst Clay.

Ardingly Sandstone was again observed in old pits [54274580 and 54404583] 500 to 700 yd E. of Old Park House, formerly Park House, west of Leigh. Examination of the sand from this locality indicated that the average grain size was between 0·15 and 0·36 mm while the largest grain was 0·45 mm. In the roadside bank nearby [54434588], some 200 yd S. of Paul's Farm, the basal shale of the Grinstead Clay was seen to rest on 2 ft of cross-bedded sandstone which in turn overlie thinly bedded sandstone.

North of the main Redleaf outcrop smaller outcrops are to be found in the wood (Redleaf Wood, 525458) 300 yd S.E. of Moorden Farm, the old quarry opposite Moorden (see Plate IVA), in the steep bank [520458] 350 yd S.W. of Moorden and as isolated crags protruding through the soil cover at 518459 and 51584608. The old quarry 100 yd S. of Sandhole [509459] must have formerly exposed the Grinstead Clay/Ardingly Sandstone junction, but now only 3 to 4 ft of massive sandstone are exposed. Berry (1961) described a section from this pit but mistook the Ardingly Sandstone for Upper Tunbridge Wells Sand.

(b) Southern limb of the Penshurst–Gilridge Anticline.

East of Wickens, 800 yd E.S.E. of Cowden Station, is an extensive exposure of Ardingly Sandstone, part of which lies on the adjoining Tunbridge Wells (303) Sheet. Crags of sandstone can be traced all the way round the two-mile outcrop, but are best developed along the northern edge of the outlier. The Lower Grinstead Clay and Cuckfield Stone cap the higher part of the hill. The total thickness of the Lower Tunbridge Wells Sand is here about 120 ft, of which 50 ft can be assigned to the Ardingly Sandstone. This outcrop is separated by a small stream from a much larger outlier to the east. Exposures of the Ardingly Sandstone can be found all round the outcrop but the most impressive crags occur to the north-west. By the roadside at Hoath House [491427] 18 ft of massive cross-bedded sandrock overlie 12 ft of evenly bedded sandstone (Plate IIIB). Sandrock weathering locally into sand has been cut by the road running north from Chiddingstone Hoath and is well exposed in the old quarry [49704305] 300 yd E. of Trugger's Farm; a sample from this locality showed an average grain-size between 0·04 and 0·15 mm with a maximum of 0·54 mm and an iron content of 0·035 per cent. Sandrock crags are developed just north of Stonewall [50304245] and massive sandstone was formerly quarried at the farm (Brookers' Farm, 49704232) 600 yd W. of Stonewall. The fault plane of a small east–west fault is well displayed in these pits.

Limestone and calcareous sandstone in association with evenly bedded sandstone at a horizon about the middle of the formation were noted in a small quarry 450 yd S.W. of the inn at Coldharbour [50154178]. In a much overgrown quarry [52154160] by the roadside 350 yd S.W. of Saint's Hill, were recorded 10 ft of white flaggy and massive sandstone, locally ferruginous. A loose block of massive calcareous sandstone was found in the stream (Frienden Gill, 50194165) 600 yd W.N.W. of Finch Green. The contributory bed was not found.

In the village of Speldhurst sandrock and massive sandstone, interbedded with some thin flaggy sandstone and much disturbed by cambering, crop out by the roadside [55554155] between 100 and 300 yd N.E. of St. Mary's Church. Similar flaggy beds, with thin clay seams, were seen to advantage south-west of the village [54704067] a little beyond the map area (Plate IIIA). Near the base of the formation the beds are folded into a slight syncline.

Eastward of Leigh, along the northern bank of the Medway, there are several overgrown pits from which the Ardingly Sandstone has been quarried. The junction with Grinstead Clay passes through the top of some of them. East of the Powder Mills the Ardingly Sandstone dies out as a mappable horizon. Massive sandstone was again noted in an old quarry [57904655] 100 yd N.W. of Barden Park, Tonbridge; in the railway cutting between Barden Park and Tonbridge ferruginous soft yellow sand occurs and sandstone from near the base of the formation was exposed during street excavations south of the railway line, north-west of Brook House.

The Quarry Hill brickworks [587449] has long been famous for the 'reversed' fault held to thrust Ashdown Beds from the south over Wadhurst Clay to the north (Plate VA). The 'Ashdown' Beds are now known to be Lower Tunbridge Wells Sand. The mistake first appears on the Old Series One-inch sheet 6 (Bromley) surveyed by Drew in the late 1850's. No fault is shown on the original map (1864) and his interpretation of the sand at this horizon, as Ashdown Beds, is undoubtedly the result of the correct interpretation of Ashdown Beds faulted against Wadhurst Clay in the railway cutting ½ mile E. by Prestwich and Morris (1846, pp. 399, 400). However, the railway cutting fault is only one of a complex of normal faults, similar to the Quarry Hill fault in that the fault planes dip steeply southwards. The junction of the Lower Tunbridge Wells Sand and Wadhurst Clay is shown by the mapping evidence to be conformable in the fields to the south of Quarry Hill. All later references to the reversed fault appear to be based on Drew's earlier misinterpretation.

Excavations in the Great Bounds Estate [57354340] proved evenly bedded sandstones, locally weathered to sand and with thin ferruginous layers. Low crags of sandrock crop out along the western side of the recreation ground [56554330] immediately north of Bidborough church.

A section noted on a western face of the High Brooms clay pit (see p. 41) showed hard and soft sandrock in beds up to 4 ft thick. This overlies Wadhurst Clay. On an eastern face similar beds up to 5 ft thick were present, and near the base was a 3-in bed of ferruginous sandstone, overlying a 3-in band of lignite, this underlain by 9 in of massive sandrock with lignite, resting on the Wadhurst Clay. Cambering of the sandstone beds increases towards the valley side in this pit.

By the side of the road from High Brooms to Old Forge several old pits, now largely overgrown or built over, showed sand and massive and flaggy sandstones, the following section from one of these pits [58904175] 650 yd N.W. of Southborough railway station being typical: buff sandy thinly bedded clay with pale orange cubically jointed sandstone, 4 ft; blue shaly clay, 2 ft; massive sandstone, white with orange specks weathering yellow, 2 ft 2 in; clay, 2 in; orange sandstone with thin clay partings and with black infiltration along bedding planes (locally forms lenses), 2 ft 2 in; orange, brown and blue clay, 3 in; thinly bedded sandstone, weathering into small rectangular blocks and interbedded with thin clay layers, 2 ft 6 in; thinly bedded sandstone with clay interlaminations and plant remains, seen to 4 ft.

(A. 10274)

A.   Joints in Ardingly Sandstone; Moorden Farm, near Penshurst Station

PLATE IV

B.   The Chiding Stone, Chiddingstone

(A. 10275)

Pale yellow sandrock extends to the base of the Lower Tunbridge Wells Sand in the railway cutting [592456] 200 yd N. of the northern end of the tunnel just south of Tonbridge. Up to 40 ft of the sand and sandrock are visible in the cutting ¼ mile N.W. of Priory Mill [599460]. Similar rock, interbedded with a subordinate amount of grey and buff sand and thin bands of grey clay, was noted in the cuttings north-west and north of Tudeley. Again in this area a roadside quarry [608461], 150 yd S.E. of Postern, showed 3 ft of thinly bedded sandstone overlying 11 ft of pale brown massive sandstone, the latter giving rise to a feature traceable eastwards for over ½ mile and exposed beneath brickearth in the roadside pits [62054585 and 62114597] 200 yd E. of Sandling Farm.

At Capel, massive white sandstone weathering to sand and associated with thin blue clay was found in an old pit [63254540] 750 yd N.W. of Tatlingbury. The sandstone forms a feature along the margin of the brickearth as far as the inn some 400 yd W. of Tatlingbury. Sand and sandstone with subordinate clay were noted in stream sections and road banks north of Capel.

In a quarry [61364460] 500 yd S. 30° W. of Park Farm the following section was recorded by Dr. Welch: yellowish buff soft sand with traces of bedding, 4 ft; yellowish buff flaggy sandstone (beds up to 2 in), 2 ft; yellowish buff soft sandrock, 2½ ft; coarser cross-bedded sandstone with irony partings, 2 in; buff cross-bedded sandrock, 1½ ft; yellow cross-bedded sandrock with irony partings, 1½ ft; buff cross-bedded sandrock, seen to 2 ft.

About a mile E. of Southborough the railway cuts some 20 ft of Lower Tunbridge Wells Sand overlying 15 ft of Wadhurst Clay. The section is no longer exposed but was recorded by Prestwich and Morris (1846, p. 401) as follows:

3. Yellow clay and imperfect sandstone.
    Whitish sand and clay.
2. Soft yellow and ferruginous sandstones only occasionally massive, with black partings, mixed and unbedded in reddish clay, with small concretionary fragments of ironstone. The lower part of this stratum at the south end of the cutting passes into red clay and small ironstones only, with an underlie of grey clay full of ferruginous matter.
1. Shales and clays. South end—upper part green, then whitish passing down into dark grey clay. As this bed trends northward, all the upper part becomes of a dark grey, passing down into a light greenish grey. Contains thin sandstone slabs covered with vermiform impressions, but no organic remains.

The top of the Wadhurst Clay is taken as the red clay in Bed 2.

A section in an old brickfield 700 yd E.S.E. of the Convent, formerly Broom Hill [57504142], and south of the Broom Hill fault, showed 10 ft of sandstone immediately under the Grinstead Clay (p. 47). This pit is now used as a council rubbish tip and is largely filled in. In the immediate vicinity of Pembury Hospital, formerly the Tonbridge Workhouse [614414], 3½ miles S.S.E. of Tonbridge there are several quarries in which sandrock predominates. The section [61254145] just west of the road fork was recorded by Dr. Welch in 1931 as (in downward sequence): clayey sand, 6 ft; massive sandstone, 10 ft; blue clay seen at south end rising to the north and dying out in the centre of the face and in the north part of the quarry represented by clay partings marked by a line of seepages, 0 to 1 ft 2 in; massive sandrock, locally soft with irony veins and dug for sand, 7 ft.

Some 250 yd N.E. of the road fork a further section [61454160] showed, in downward sequence: clayey sand, 2 ft; hard sandstone, thinly bedded in 4-in jointed layers, 1 ft 10 in; reddish clayey sand, 1 ft; massive sandstone, 1 ft; soft clayey sandstone, 6 in; clay and sandy clay, 6 in; soft clayey sandstone, 2 ft 6 in; soft cream-coloured calcareous sandstone, 9 in; soft cream-coloured calcareous clayey sandstone, 6 ft; orange-coloured ferruginous sandstone, 9 in; massive jointed sandstone, seen to 16 ft.

Allen (1959, p. 298; 1962, p. 241) redescribed the Pembury Wood sandpit in the light of his continuing researches into Wealden palaeogeography, and compared it with the section exposed in the fork of the roads opposite the hospital (Blackhurst Lane section, 613414). The Blackhurst Lane section was regarded as the 'normal' local succession. The important horizons noted were S.W.-building conelets (channel mouth-bar?) seven to eleven feet high at the base of the section and a rootlet marker-bed and massive sandstone (crevasse sheets?) above. In the section in the Pembury Wood pit a large channel trending S.S.E. cut out everything above the rootlet bed. The channel was filled with clays, silts and lenticular sandstones, and ran nearly parallel to the underlying foresets. Allen suggested that deposition of the lower sands and cutting of the major channel probably took place contemporaneously with delta-growth in the south; the subsequent silting of the channel may have constituted a local response to the distant rise of lake level (Grinstead transgression).

Numerous small exposures of soft sand, and several of orange coloured and white sandrock and sandstone occur in Pembury Wood. North of Lower Green, by the roadside [62754230] 250 yd S. of the Water Works, massive sandstone low down in the succession is broken by cambering, while widening of the east–west road at Cott's Hill crossroads [64904385] revealed minor folds in sandrock.

The Tunbridge Wells Sand is broken up into a number of small outliers between Crockhurst Street [624448] and Kenward [62354350]. Some, such as those of Knowles Bank [62204405], cap low hills while others are low down in the valley bottoms as a result of cambering (the outcrops 400 yd W. and N.W. of Knowles Bank), or owing to faulting (the outcrop [635435] 500 yd S.E. of Bouncers Bank). Dips of 30° N.N.E. were recorded in the stream (Devils Gill, 61644408) where thick and flaggy iron-stained sandstone was exposed. A similar section occurs in the stream [61824410] 200 yd E.

Massive sandrock dipping 7° N.E. was formerly quarried in the old pit [62604459] 400 yd S.S.E. of Crockhurst Street Farm. The small stream west of Reeds Farm is floored by massive sandrock dipping 20° N. at 63994374. This outcrop is thrown down by a N.E.–S.W. fault against Wadhurst Clay; this clay can be augered in the fields 40 yd S.E. of the sandrock outcrop.

Thin beds of sandstone noted along the road between Crittenden and Five Wents, and massive sandstone by the roadside [66434265] 350 yd W.S.W. of Crundalls Farm, are horizontally bedded.

Near the crossroads [67954225] 550 yd N. of All Saints' Church, Brenchley, beds of sandstone and sandrock dip south; a southerly dip is evident in the valley just south of Brenchley.

A 10-ft waterfall is formed by thickly bedded sandstone at 676414, 500 yd S.W. of the church. Similar beds crop out in the bed of the stream 500 yd downstream [68074145].

C.R.B., S.B.

GRINSTEAD CLAY AND CUCKFIELD STONE

South and east of Ford Manor, near Dormans Land, a belt of heavy clay land marks the outcrop of the Grinstead Clay, in which there are many old overgrown and flooded pits.

A temporary shallow trench dug from Greybury [43804265], about 2 miles S.S.W. of Edenbridge, in a north-westerly direction traversed the whole of the Grinstead Clay outcrop and showed the lowest beds of dark buff clay to be succeeded upwards by two beds of fine sandstone separated by loam. Above these, loamy clay passes up to white and then to reddish brown clay. Stiff dark grey clay follows. The highest bed is pale grey clay.

A small faulted inlier can be found by the Mill [696459] 900 yd N.W. of Chidding-stone church. Numerous overgrown pits mark its outcrop, and it is overlain at one point [49644592] by a thin outlier of Upper Tunbridge Wells Sand. North of the River Eden a line of old pits near Sandhole [509460] pick out the Grinstead Clay outcrop. There is a narrow strip of Ardingly Sandstone between the clay and the first terrace of the River Eden (see p. 43). To the north, although largely obscured by Drift deposits, the Grinstead Clay is faulted against the Weald Clay. East of Chiddingstone church grey shaly clay can still be found in the sides of the old pit at 50404512.

A group of old clay pits in the former grounds of Redleaf House (The Grove [520457], Redleaf Wood [527459]) mark the Grinstead Clay outcrop. To the east of the house the clay thins and is overlain by the Upper Tunbridge Wells Sand. A section in the road [52504522] 200 yd E. of the house showed grey shaly clay overlying thick iron-stained sandstone dipping 3° N. 30° E. Farther east, to the south of Leigh, and separated by a small fault from the Redleaf outlier, the Grinstead Clay is picked out by a further line of old pits. A road cutting [54424593] 200 yd S. of Paul's Farm reveals 10 ft of grey shaly clay, while 30 yd S. of this exposure the junction of the Grinstead Clay and Ardingly Sandstone was observed. The dip of the sandstone was 8° N.

The clay to the north of Great Barnetts, formerly Barnetts, has become silty and changed to a red colour. East and north of this locality the Grinstead Clay has not been found.

The Finch Green outlier to the south-west of Penshurst is preserved in a slightly asymmetrical syncline, of which only the northern limb is seen in the map area. In the centre of the syncline a small remnant of Upper Tunbridge Wells Sand is preserved. The Upper and Lower Grinstead Clay are about 20 ft thick while the Cuckfield Stone is slightly thicker, averaging 25 to 30 ft. Old pits everywhere mark the outcrop of the Lower Grinstead Clay. Massive Ardingly Sandstone is exposed beneath slipped grey clay in the old pit [51104203] 300 yd N.W. of Coldharbour. Ironstone nodules were found on the surface of the field [507418] 200 yd N. of Finch Green but the contributory bed was not seen *in situ*. The top of the Lower Grinstead Clay is invariably reddened where overlain by the Cuckfield Stone. The Cuckfield Stone has a large dip-slope outcrop and fragments of sandstone are usually abundant on its surface. Calcareous sandstone was found at one point [49524210]. Thin flaggy greyish yellow sandstone is exposed in the road bank [50154229] 200 yd S.W. of Stonewall.

The Upper Grinstead Clay has a very restricted outcrop on the southern margin of the map area. Lithologically it is identical to the lower clay.

The Southborough outcrop is fault bounded on the north. It also contains a number of old pits, in one of which—an abandoned brickfield [575414] 700 yd E.S.E. of Broom Hill—the following section of Grinstead Clay was seen: ferruginous shale, 6 in; yellow and buff shale, 2 ft; rock marl, 9 in; yellow and buff shale, 2 ft; rock marl, 1 ft; buff, yellow and grey shale with nodules of rock marl, 3 ft; rock marl with purple facings, 1 ft; buff and grey calcareous shale, 2 ft; purple and grey mottled rock marl, 9 in; buff shale, 1 ft 8 in; calcareous shale, 4 in. It rests on 10 ft of massive sandstone of the Lower Tunbridge Wells Sand.

In the Lower Green outlier the following section was seen by Dr. Welch in 1931 at the Pembury brick and tile works: Upper Tunbridge Wells Sand comprising reddish sandy soil, 1 ft 9 in on soft orange sandstone, locally white, with irony veins, 6 ft on hard ferruginous sandstone, 0 to 5 ft; Grinstead Clay consisting of blue laminated clay with mottled black and ochreous streaky layers and some grey and red mottled clay, up to 22 ft on orange sandy shaly clay, 1½ ft; Lower Tunbridge Wells Sand, hard massive well-bedded and jointed fine-grained pale yellow sandstone, seen to 3 ft.   C.R.B., S.B.

UPPER TUNBRIDGE WELLS SAND

The deposits in the extreme north-west corner of the map area are probably quite high in the succession and consist of fine silts and silty clays.

In and around Dormans Land several overgrown pits, from which sandstone formerly had been worked, and numerous other small exposures, all show thinly bedded, grey and yellow, fine-grained, soft sandstone or siltstone associated with sand, loam and loamy clay. In general the beds dip to the north-west, but gentle folding has been noted, e.g. in the roadside bank [40744183] 400 yd N.E. of Farindons, north-east of Dormans Land. Thin flaggy sandstone was also seen on the roadside banks [42754290] of the lane leading to Hoopers, formerly Hopers, Farm and again in old pits [43184313] immediately south of Newbarns Farm. The dip was 55° N.W. at the latter locality.

The Upper Tunbridge Wells Sand of the Redleaf outcrop to the east is composed of loam, soft sand and thinly bedded sandstone, occasionally ferruginous. These same beds were noted in the railway cutting [535463] to the north. Sandrock was also seen in a temporary excavation [54554655] 200 yd S. of Hall Place, Leigh, and flaggy sandstone in the churchyard [54834656] 400 yd E.   C.R.B., S.B.

WEALD CLAY

Around Waterside [395452], 1 mile N. of Lingfield Station, the Weald Clay is mottled grey and buff, weathering uniformly buff. Shallow temporary exposures east of the railway revealed shale under clay and occasional thin layers of fine-grained sandstone or siltstone. Thin siltstone fragments were found in the clay around shallow pits south of Edenbridge but no exposures were seen of siltstone *in situ.* Red and grey mottled clay crops out at How Green some 800 yd N.N.W. of Hever Castle.

In the railway cutting east of Penshurst Station the lower beds of the Weald Clay consist of bluish grey and pale grey shale weathering to bluish grey and almost white clay, with thin yellow siltstone interbedded.

Overgrown and almost obliterated clay pits south of Paddock Wood marked the sites of former brick and tile works, [675449, 67684468, 678446 and 674436]. West of Mascall's Court thick sandstone is exposed in the road bank [67554370]. This sandstone can only be traced for a short distance in the fields to the north and north-east as it is overlain by Head in this direction.

About 1¾ miles S. of Limpsfield, in the pit of the Red Lane brick works [407502], 6 in of loam with small angular flints and pebbles were seen to overlie mottled buff-grey stiff clay 7 to 10 ft thick in which is a seam 4 to 12 in thick of buff sandstone. This rests on reddish brown clay, seen for 3 ft.

Over much of the area north of Edenbridge buff or yellowish brown clay seen at the surface passes down into grey and buff mottled clay; in the railway cuttings in the vicinity of the crossing [426477] 1¾ miles N.W. of Edenbridge bluish grey shale, which underlies the above, was seen. Brown shale associated with grey loam was exposed in the railway cutting [485469] ¼ mile W. of the bridge at Bough Beech, and shales in stream sections north-east of Chiddingstone.

Around Hildenborough the clay is stiff and usually mottled yellow, buff and grey, but north-east of Tonbridge a considerable area of the Weald Clay is covered by a re-sorted loamy material.

In the vicinity of Itchingwood Common, 1½ miles S.E. of Limpsfield, the clay is stiff and grey in colour and on weathering shows a greyish brown, yellow mottling. At the eastern end of the Common a roadside section [42075067] showed pale grey and ochreous clay, while reddish brown clay was seen in a roadside trench [42255070] between the Common and Moat Farm, and again 150 yd S. of the farm.

Irony sand and clay, pale grey clay and red and grey mottled clay in descending order occur in the road bank [446500] 350 yd W. of Crockham Grange, to the east of which the clay is usually buff or mottled brown and grey, locally weathered yellow. Reddish brown clay was again observed 250 to 450 yd S. of Chain Farm, formerly Chains [49005095], near Ide Hill, and at the southern end of the Sevenoaks railway tunnel.

Yellow shale associated with siltstone and overlain by white clay and siltstone is exposed in the stream section [55945132] ¼ mile N.W. of Tumbling Bay. A little eastwards [56925150] white clay occurs between a lower sandstone and a limestone some 600 yd N.W. of Fairhill. Between a higher sandstone and this limestone, the shale was seen to be gently folded [572518].

Stiff pale grey and ochreous mottled clay persists around Shipbourne except 750 yd N.W. of the church where the clay is white [58505256], and 700 yd W.S.W. of the church where it is reddish brown and grey mottled [58565196]. Reddish brown and grey mottled clay was also noted 170 yd S.S.E. of Hookwood [60805133].

Sandstone and limestone beds in the Weald Clay. Outcrops of these subordinate beds may not be traced far across the Weald, partly owing to the lenticular nature of the beds, and partly to the paucity of exposures. The known occurrences are described below in three groups, from west to east, commencing with the lowest group. On the one-inch and six-inch sheets a conventional representation of these outcrops of thin sandstones is in general all that has been attempted.

Topley (1875, p. 105) mentions that his No. 3 limestone was dug for roads around Bowerland Farm, 3 miles W. of Edenbridge. East of the farm are numerous abandoned shallow pits in many of which fragments of Small-'*Paludina*' limestone are to be found, and in the fields limestone fragments occur on the surface; in only one pit [397463], 500 yd N.N.E. of the farm, was the limestone seen *in situ*. Over this area thin buff siltstone beds were noted in temporary excavations but their outcrops could not be traced.

South of Edenbridge, along a ridge which marks the strike of the beds east from the 27th milestone [44554545], fragments of dense, grey, unfossiliferous limestone have been noted, but no outcrop of limestone was found in the several old pits along this feature. Many of these fragments are surrounded by a yellow crust, ⅛ in thick, of fine-grained non-calcareous silt resembling Tunbridge Wells Sand and probably the result of leaching out of lime; this suggests that some of the thin sandstone layers seen elsewhere may in fact be decalcified beds.

Thin slabs of limestone up to 1½ in thick crowded with *Filosina* and *Viviparus* were seen under gravel in a temporary section [44304614] on the Lingfield Road, 100 yd W. of High Street, Edenbridge; sand was noted some 700 yd E. of Syliards [48104754], and slabs of limestone in the stream bed nearby [48004746]. A shelly limestone, separated by clay above and below from beds of fine-grained sandstone, crops out in the stream bed [49284742] about 350 yd S.S.E. of Bough-Beech Place, formerly Ivy House, now demolished; sandstone also occurs in the stream bed [49454779] 300 yd E.N.E. of this site, and was exposed in an old pit 750 yd N.E., while another limestone, separated by clay from sandstone above and below, crops out in the stream bed 650 yd N.N.E. The lower sandstone of this last group was traced across the fields for about ¼ mile E.

In 1967, a trench [49094751 to 49144747] excavated along the centre line of the Bough Beech Reservoir dam revealed contorted siltstones, sandstone, limestones and clay. The effects of valley bulging appear to be confined to the western side of the valley. The following is a composite section recorded by Dr. R. G. Thurrell:

	Ft	in
Clay, silty, yellowish brown mottled light greenish grey, deeply weathered at western end of pit seen to	6	0

Ft in

Sandstone, fine-grained silty, compact calcareous, pale bluish grey,
weathering buff or yellowish brown. Fissile cross-bedded on small
scale, with stick-like bottom markings tending to be lenticular
and with incipient formation of basin casts .. .. 2 in to 1 0

Clay, yellowish brown, tending to be very silty in some places,
otherwise pale grey; approx. 2 6

Pebble bed: gritty medium-grained sandstone, hard, calcareous,
pyritic, cross-bedded, dark bluish grey to brownish grey, very
impersistent (maximum) 0 2

Clay or shaly clay, unctuous, greyish green; from .. 4 ft to 5 0

Limestone: dark bluish grey, dense shell fragmental limestone, with
occasional whole shells of small *Viviparus* (Small-'*Paludina*'
limestone) up to 0 4½

Clay, slightly silty, homogeneous, sticky, with occasional bands
ostracod-bearing, dark grey to greyish black 3 0

Siltstone, weathering to silt and clayey silt, much broken, deeply
weathered. Pale greyish to yellowish brown to reddish brown,
ferruginized. Well bedded, fissile but no microfossils seen: thickness variable from .. .. 2 ft to 3 0

Clay, silty, striped, with sporadic gritty bands up to 2 in thick, bluish
grey to grey. Strongly striped 3½ ft below top. Basal 3–5 ft
weathered to a distinctive yellow and brown silty mudstone more
silty than rest, well bedded, with a closely spaced 'blocky'
fracture .. .. 10ft to 15 0

Clay, very shaly, dark bluish grey to dark grey, homogeneous,
occasional ostracods seen; contains (1) limestone from 3 to 5 ft
below top, hard, dense, shell-fragmental, pyritic with scattered
bones and denticles, but consisting mainly of freshwater gastropods (Small-'*Paludina*' limestone) from 3 to 6 in thick; (2) clay
ironstone band, stratigraphically from 3 to 4 ft below the limestone, reverts sporadically to non-ferruginous siltstone of pale
cream to buff colour, 1–3 in thick. At the eastern end of the
excavation this incipient ironstone is associated with a calcareous
siltstone between 6 and 12 in below it; dark grey in colour,
strongly cross bedded and fissile; up to 12 in thick but variable.
Thickness seen approx. 15 0

Total thickness of strata recorded is approximately 45 ft.

In the same district a limestone associated with a sandstone occurs in the bed of
the stream [50584780] 700 yd S.E. of Kilnhouse Farm, a sandstone in the stream
700 yd E. [50724815], and a Small-'*Paludina*' limestone 750 yd E.N.E. [50794836];
sandstone and limestone fragments were noted [51684875] 350 yd E. of Sharp's Place;
and limestone in the stream [51654801] 150 yd N. of Brownings, 400 to 700 yd N.E.
of which fragments of sandstone were observed around the edge of a number of
abandoned, waterlogged pits.

North-east of Chiddingstone, in the immediate vicinity of Wickhurst [527479],
limestone occurs between horizons of sandstone, while in the stream [52804872]
600 yd W. of Southwood sandstone and sandy shale were exposed. Fine-grained
sandstone fragments were common 100 yd downstream while 100 yd upstream were
many slabs of Large-'*Paludina*' limestone.

Topley (1875, p. 99) gives the following section from the railway cutting about
¼ mile N.W. from Hildenborough Station, where he recorded a dip of 5°, in a direction
35° W. of N: clay and shale, 20 ft; shelly limestone, 6 in; shale and sandy clay
(?calcareous), 8 ft; shelly limestone with 'beef' at bottom, 1 in; shale, 5 ft; shelly

limestone with *Cyrena* [=*Filosina*], 1 to 1½ in; shale (*Cypridea* and *Cyrena*), 5 in; shelly limestone with 'beef' at top, ¾ in; shale, 6 ft. Crystals of selenite were very common.

A short distance farther north in the cutting some 200 yd S.E. of Philpots abundant fragments of fine-grained sandstone suggested the proximity of a sandstone bed.

In the abandoned pit in a brickworks [598495] 450 yd S.W. of Starvecrow, 2 miles N.N.E. of Tonbridge, there are thin bands of blue limestone with Small-'*Paludina*', *Cassiope* and *Filosina*. The Small-'*Paludina*' limestone was in two levels, about 10 ft apart, above two beds of '*Cyrena*' limestone, also about 10 ft apart, the lower yielding *Cassiope strombiformis* (Schlotheim) (see p. 34 and Abbott 1907b, p. 100). '*Cyrena*' in limestone beds underlying Small-'*Paludina*' limestone occurs at the base of Topley's Bed 3 in the Maidstone district, in the Warlingham borehole and in the Horsham district (Worssam and Thurrell 1967).

Near the western margin of the area, about ¼ mile S. of Oldhouse Farm, soft buff micaceous sandstone has been worked in a shallow pit [402470], and a small pit [393474] about 1000 yd W. of Oldhouse Farm showed, beneath a foot of brown loamy clay soil, 1½ to 2 ft of brown to orange earthy loam with rubbly sandstone underlain by buff, yellow and pink sandstone, fine-grained and with shell casts, divided into layers 4 to 6 in thick by 1-in bands of irony sand, seen for 2½ ft.

About ¾ mile N., and immediately west of Sunt Farm, ditches cut on either side of the railway cutting [401486] exposed clayey sand, sand and buff and brown sandstone, and sandstone was encountered in foundations for electricity pylons alongside the cutting.

Between Whitehouse Farm and the Holland–Haxted road a shallow pit [41064823] exposed 3 ft of buff sandstone with ferruginous partings, covered by 1 to 2 ft of dark brown sandy soil.

From the railway cutting [422487] W.N.W. of Batchelor's Farm, Topley (1875, p. 106) recorded sand and sandstone of his No. 5 horizon, 20 ft thick at least, dipping at the northern end under clay at 3° or 4° N. At the southern end of the cutting the dip is given as 6° N. Soft buff or brown, iron-stained sandstone was still to be seen in 1946. Clayey sand was noted on the southern slope of an abandoned pit [42884867] some 350 yd N.E. of Batchelor's Farm; a topographical feature immediately east of this pit may indicate the presence of sandstone. Soft yellowish buff, fine-grained sand occurs at the southern end of a pit [44004875] 450 yd N.E., and soft yellowish brown sandstone was seen in a temporary exposure on the roadside 150 yd S.S.E. of Lyndhurst.

A belt within which there is considerable evidence of more than one sandstone bed extends from Winkhurst Green, 2¼ miles S.W. of Sevenoaks Weald, to Hollanden House, ¾ mile N.W. of Hildenborough. Within this belt limestone was observed at only two points, in an old pit [505493] 350 yd N. of Bore Place where it was associated with sandstone, and in the stream bed [52754933] 900 yd N.W. of Southwood where it contained *Viviparus*. Sandstone fragments were found in clay pits [509494, 51054954] 550 yd W.S.W., 350 yd W., and at several points around the boundaries of fields ¼ mile E. of Bushes Farm, while a 3-in thick bed of sandstone was exposed in an old pit [51734980] 500 yd N.E. of the farm. Farther east, in the stream bed about 100 yd above the outcrop of Small-'*Paludina*' limestone near Southwood, sandstone fragments were sufficiently abundant to suggest the proximity of a sandstone bed. Still higher upstream [52734967] ¼ mile S.E. of New House is a thin bed of sandstone, and above that point in the stream [52774990] 300 yd E.S.E. of New House a sandstone

E

was seen to be at least 18 in thick, but the full thickness of the bed was not determined. Sandstone was found in an excavation at Westwood [529495], 700 yd S.E. of New House where it is fine grained and occurs as thin slabs. East of the Hildenborough–Sevenoaks road, 900 yd E.N.E. of Mansers, sandstone was also observed [55504983] and again in temporary sections 200 yd and 300 yd S. and 250 yd S.E. of Hollanden House [55954975].

About 1¾ miles S. of Limpsfield, in the pit of the Red Lane brick works [407502], occurs a seam of weathered buff sandstone locally speckled green, 4 to 12 in thick. Almost a mile farther north soft buff sandstone passing down into soft medium-grained sand near the top of the Weald Clay was dug east of the former Greenhurst Park over a distance of about 200 ft by the side of the road [40705175] between Limpsfield and Foyleriding Farm. This sand is higher in the Weald Clay than Topley's horizon No. 7 and is not elsewhere exposed.

South-west of Crockham Hill yellow sand and soft sandstone were exposed by the roadside [43705025] 400 yd E. of Hurst Farm, and similar sand was noted by the roadside [44684998] 350 yd W. of Crockham Grange.

A bed of limestone 1 ft thick containing large *Viviparus* occurs 450 yd S.S.E. of St. George's Church [53075096] at Sevenoaks Weald. Opposite the smithy this bed is said locally to become soft on weathering, and also in places to form two layers. The conjectural position of the limestone outcrop follows a south-westerly direction for a distance of over a mile along a topographical feature and runs through numerous abandoned shallow pits from which limestone was probably once quarried. This outcrop is the continuation of Topley's Bed 6 (the Large-'*Paludina*') limestone which he recorded north of Elses, formerly Elsey's, Farm and from the Sevenoaks and Tunbridge railway where he recorded a thickness of 22 in, sometimes as one bed, sometimes as two.

Eastwards from St. Julian's Farm, formerly Ramstead Farm, [551518], a well-marked feature and a number of old pits mark another probable limestone horizon; 550 yd E. of Goldings a layer of limestone 4½ in thick was exposed in the bed of the stream [57075170]; elsewhere scattered fragments of Large-'*Paludina*' limestone occur. This may be an extension of Topley's Bed 6 horizon recorded a little farther east at Budd's Green. About 100 yd N. of the last exposure [571518], the section in downward sequence is ferruginous sandstone 6 in, yellow and grey mottled clay 6 in, ferruginous sandstone 2 in, clay 4 in, ferruginous sandstone 2 in, resting on shale.

Several thin siltstone beds crop out in a stream section [55935133] 400 yd N.W. of Tumbling Bay. East of Fairhill [574512] a belt of loamy soil with fragments of sand-stone 500 to 600 yd wide extends to the Tonbridge–Shipbourne road. Sand was exposed 250 yd W. [58095116]; sandstone in an old pit 500 yd S.W. [580508]; loam with sandstone fragments 350 yd S.S.W. [58235084] and ferruginous sandstone in old pits [58655078, 58675062] 550 yd and 700 yd S.E. of Tinley Lodge; loam with fer-ruginous sand also occurs 1300 yd S.E. [59345056] of this point. The soil and vegeta-tion, and the frequency with which old pits occur along this belt suggest the presence of several sandstone beds.

Limestone was recorded by Topley (1875, p. 107) as having been exposed in drains on Hadlow Common, due west of Goose Green.

The highest beds of the Weald Clay were exposed during the excavations for the Sevenoaks–Tonbridge railway line (Caleb Evans 1871, pp. 1–3). Slightly annotated the section runs:

Atherfield Clay

e. Dark greyish coloured sandy clay which gradually passes, by the loss of argillaceous matter, into a dark clayey sand. Large concreted masses at this horizon contained abundant shells identified as *Perna mulleti, Corbula striatula, Terebratula sella,* etc.

Weald Clay

d. Dark, almost black clay containing *Cerithium* or *Potamides, Ostrea, Cardium.*

c. Green, buff and brown clays somewhat shaly. Beds and concretionary masses up to 2 to 3 in thick composed almost entirely of *Cyrena* are common in the upper layers. In some blocks are layers of a small *Paludina; Cypridea tuberculata* covers the surfaces of many of the blocks; *Lepidotus* and turtle have been found. Beds of *Unio* are present in the lower layers.

b. Large-'*Paludina*' limestone. Hard Sussex marble composed almost entirely of the shells of *Paludina* and with occasional *Cyrena.* 2 ft thick.

a. Grey, bluish and greenish clays with many layers of *Unio* and *Cyrena.*

Unfortunately with the exception of Bed b. no thicknesses are given. Bed b. lay at a depth of 127 ft from the surface in Shaft 3. This is presumably the 3rd shaft [53395206] working northwards from the southern end of the tunnel. On this assumption the Large-'*Paludina*' limestone lies at a depth of about 150 ft from the base of the Atherfield Clay, a figure which is in accord with that deduced from surface outcrop nearby.

S.B., C.R.B.

Chapter IV

CRETACEOUS: LOWER GREENSAND, GAULT AND UPPER GREENSAND

GENERAL ACCOUNT

LOWER GREENSAND

CONDITIONS UNDER which the fresh and brackish water series of Wealden strata were laid down were terminated by an incursion of sea into the Wealden area. This resulted in the deposition of a series of marine argillaceous and arenaceous sediments known as the Lower Greensand, separable on lithological characters into four subdivisions, namely, in upward sequence, the Atherfield Clay, the Hythe Beds, the Sandgate Beds and the Folkestone Beds. These are recognizable almost throughout the Lower Greensand outcrop of the Weald and are all represented in the present district.

The Atherfield Clay succeeds the Weald Clay with only slight changes in lithology, but at its base the establishment of marine conditions is shown by a fossil marine fauna. This bed is but rarely to be seen. The remainder of the Lower Greensand is essentially arenaceous, in which changes in lithological type, both laterally and vertically, are frequent along the northern side of the Weald, suggesting shallow littoral conditions. The Hythe Beds, sandy in the main, contain layers of calcareous sandstone or sandy limestone (the Kentish Rag), which, apart from a development of chert beds (the Sevenoaks Stone) at the top of the formation, occur at intervals throughout the formation, but are more common in the east of the present district than in the west. The Sandgate Beds contain a high proportion of glauconite which imparts a bright green colour to the deposits when fresh, though they weather to a bright red loam; this formation also contains some clay beds locally which render it impervious. The Folkestone Beds are generally false-bedded more or less ferruginous sands, which show little variation except that in the upper part between Seal and Ightham certain hard beds, including a cherty facies, resemble those at the top of the Hythe Beds. A generalized section illustrating the lithological characters of the Lower Greensand formations is shown in Fig. 3, and sections across the outcrop in Fig. 4.

The Lower Greensand extends across the northern half of the district, with an outcrop of 1½ miles in width in the west, broadening to 5½ miles in the east and occupying an area of 54 square miles. The thickness of the series as a whole varies between about 250 and 450 ft. H.G.D.

In his full account of the stratigraphical palaeontology of the Lower Greensand, Casey (1961, pp. 536–46) has described and correlated a number of sections in West Kent which fall within the Sevenoaks map area. These are:

Sevenoaks railway tunnel and boreholes at Sundridge and Riverhead (Atherfield Clay, p. 537); quarries near Sundridge, Borough Green, Offham and West Malling, and a borehole at Trottiscliffe (Hythe Beds, pp. 540–1; Sandgate Beds, p. 541); sandpits near Westerham and Wrotham Heath, and a borehole at Brasted (Folkestone Beds and basal Gault, pp. 543–6); and sandpits near Ightham and Borough Green (Folkestone Beds, p. 542). S.C.A.H.

ATHERFIELD CLAY

This formation, seldom reaching a thickness of 50 ft, usually comprises reddish, yellow, grey and bluish black clays, in places a little sandy or shaly. Calcareous and ferruginous nodular claystones containing marine fossils occur at certain horizons and are frequent near the base. Its outcrop, mainly confined to the foot of the Lower Greensand escarpment, is often masked by hill-wash; it is, therefore, seldom well exposed in the field, and its lower boundary is usually not traceable with any degree of accuracy. In the present district, though precise evidence of its true thickness is wanting, there are indications that it may be 30 ft or more thick. The thickness near Maidstone, to the east of the present district, is 20 to 30 ft, and near Reigate, to the west, 30 ft.

Atherfield Clay has been recorded by Drew (in Topley 1875, p. 113) in the present district at Nettlestead on the east, as a bluish slightly sandy clay with two species of oysters and other marine fossils; in the railway tunnel at Sevenoaks (constructed 1866) where many fossils were collected by Caleb Evans (1864, p. 348; 1871, pp. 2–3; see Casey 1961, p. 537) and presented by him to the British Museum (Natural History); and at the small inlier at Whitley Mill, 2 miles S.S.W. of Sevenoaks Station, as a blue sandy clay with marine fossils. Atherfield Clay was encountered in the bottom of the railway cutting close to the northern entrance to the Sevenoaks tunnel, brought up by sharp saddle folding accompanied by faulting, and in the Hubbard's Hill vicinity during the construction of the Sevenoaks By-pass. Blocks of the 'Perna' Bed, a calcareous sandstone, were encountered in an exploratory trench [53375202]. When fresh these blocks are a greenish grey in colour with scattered glauconite grains. The surface weathers to a yellowish white but the glauconite grains appear to be unaffected. The fauna from these blocks included: Sellithyris?, Anchura [Perissoptera] robinaldina (d'Orbigny), Globularia cornueliana (d'Orbigny), Loxotoma neocomiense (d'Orbigny), Arca dupiniana d'Orbigny, Chlamys sp., umbonal fragment of large taxodont (Cucullaea?), Lopha colubrina (Lamarck), Modiolus sp., Mulletia mulleti (Deshayes), Parmicorbula striatula (J. de C. Sowerby), Resatrix sp.

Site investigation boreholes and pits for the By-pass scheme proved Atherfield Clay at a number of points westward of River Hill and showed it to be subject to broad upfolding, probably as a result of superficial movement in front of the Hythe Beds escarpment, following extensive Pleistocene erosion. The junction of the clay with overlying Hythe Beds was recorded by Mr. S. C. A. Holmes at a point [52725189] below Hubbard's Hill in a temporary section:

		Ft
Hythe Beds	{ Yellowish grey calcareous sandy loam with glauconite;	
	less friable, and with a stronger clay content, at the base	3
	Sharp but uneven junction with northerly dip up to 10°	–
Atherfield Clay	{ Chocolate to buff blocky silty clay	1
	Dark blue compact clay, more shaly; drying out to grey	6

This exposure adjoining the old road up the escarpment from Sevenoaks Weald confirmed the line of outcrop as mapped by the late H. G. Dines; it was on the only spur of ground hereabouts undisturbed by landslipping.

In a large inlier at Brook Place and Penn Farm [490530], 1½ miles S. of Sundridge, Drew found shaly clay not far below the base of the Hythe Beds and, regarding this as Weald Clay, assumed the Atherfield formation to be thin here; no fossils were recorded. Nevertheless, there is no reason to suppose that the Atherfield Clay is missing anywhere within the present district, and consequently the outcrop has been indicated on the geological map in accordance with its estimated thickness. In places along the scarp-face, especially south of Mereworth Woods and in the Shode valley, cambering (see Hollingworth and others 1944, pp. 4–11) appears to have depressed the lower boundary of the Hythe Beds. On the usual interpretation of this structure the Atherfield Clay outcrop should not be masked though it may be attenuated, but here and there small-scale landslips may cover the outcrop.

The boundary of an inlier [445525] of Atherfield Clay north of Crockhamhill Common is indicated by a line of strong seepage springs at the junction of the Atherfield Clay with the overlying permeable Hythe Beds.

Two valleys breaching the Lower Greensand escarpment near Ightham Mill, southward of Ivy Hatch, and near Wilmot Hill [580537] are also floored with Atherfield Clay, while a much larger gap occurs at the valley of the River Bourne (i.e. Shode or Plaxtol Brook, see p. 4) which cuts an almost gorge-like valley down to the Atherfield Clay throughout the whole of its course across the Hythe Beds south of a point 600 yd N. of Basted [609556], which is within 500 yd of the outcrop of the Sandgate Beds, to the north. Atherfield Clay was proved at a depth of 198 ft from ground surface in a borehole at Sundridge, and at a depth of 250 ft in a borehole at Riverhead, where the clays yielded *Deshayesites forbesi* Casey (Casey 1961, p. 537). A borehole at Mid Comp Farm [63465672], between Borough Green and Offham, reached probable Atherfield Clay at 162 ft and another [65915648] about ½ mile S. of Offham proved 42 ft of the clay beneath Hythe Beds. The formation appears to have been struck at South Green, Titsey, at about 435 ft, in a borehole of the East Surrey Water Company.   H.G.D., S.C.A.H., C.J.W.

HYTHE BEDS

The outcrop of this formation occupies the greater part of both the scarp-face and the dip-slope of the Lower Greensand outcrop as a whole, and extends right across the district, its width gradually increasing from about ¼ mile just south of Limpsfield, on the west, to about 2½ miles, south of West Malling, on the east.

Its maximum thickness is possibly a little over 200 ft in the Mereworth Woods area, though here it may be more apparent than real, for hill-creep or cambering, the magnitude of which cannot be ascertained with any degree of certainty, appears to have lowered the lower boundary as mapped on field evidence, both in the Shode valley and on the scarp-face. A borehole at Sundridge showed but 96 ft, and another at Riverhead 100½ ft. In both boreholes Sandgate Beds above, and Atherfield Clay below, were proved. Borings at the Sevenoaks Waterworks pumping station [570577], 1 mile S.E. of Kemsing, entered clayey beds at about 244 ft, but it is uncertain whether Atherfield Clay was reached; the minimum thickness of Hythe Beds present is 88 ft. At South Green, Titsey, however, 125 ft were proved before grey clay was entered.

In other localities the thickness is of the order of 150 ft, notably at Ide Hill and Bayley's Hill; it may be no more than 140 ft at West Malling, as calculated from a well record at the Abbey Brewery.

The Hythe Beds are essentially a sandy deposit, the colour being generally white, cream or pale grey, faintly tinged with green owing to scattered grains of glauconite. The sandy beds, known as 'hassock', are generally of fine or medium grade, though they are coarse at certain horizons and carry small quartz and lydite pebbles and grains of glauconite up to $\frac{1}{4}$ inch in diameter. Almost invariably there is a fair proportion of fine silty material which results in a very mild cohesion. Fossils which were originally calcareous usually retain their calcareous nature, and beds of pale bluish grey sandy limestone, the 'Kentish Rag', occur usually a foot or more thick, as well as sandstone layers with a calcareous or siliceous cement which generally consist of friable nodular masses embedded in sand. At the top of the formation come about 12 ft of greenish yellow loamy sand that weathers red, more glauconitic than the sand below it, foreshadowing the oncoming of the highly glauconitic and more clayey Sandgate Beds. At this horizon bands of honey-yellow or brown chert, often containing numerous sponge spicules and locally known as 'Sevenoaks Stone' (Topley 1875, p. 119), are common. Generally a few inches thick, but occasionally up to a foot or more, these beds may occur as continuous layers of brittle, flint-like rock or as lines of nodules with cherty cores and carious rinds. The destruction by weathering of these beds has contributed largely to the formation of the Angular Chert Drift that caps high ground of the Greensand escarpment (p. 116). A typical section of Hythe Beds is illustrated in Plate VB.

There is a gradual change within the district from a sandy lithology which is predominant in the Reigate area to the west, to a more calcareous type including the rag and hassock layers so strongly developed around Maidstone to the east. Rag and hassock are present throughout the full thickness of the formation in the eastern part of the district, but, traced westwards, the rag beds become more widely spaced, and their lime content decreases, until, in the west, they become more or less calcareous sandstones, but are still distinguishable from the hassock. Lateral changes take place also in a north–south direction, the quarries along the scarp face, mostly very old and few now active, contain fewer rag beds and thicker interbedded sand layers than the pits and cuttings on the lower parts of the dip slope near Sundridge and Ightham.

Lateral changes, both along the strike and across the dip, are also evident at the chert horizon. Chert is not well developed west of Brasted, but is present in nearly every exposure of the topmost Hythe Beds eastwards. Across the dip the thickness of the chert beds as a whole remains more or less constant, and in all exposures on the scarp crest thicknesses of individual chert bands of 1 ft or more are not uncommon; on the lower part of the dip-slope, however, they seldom exceed a few inches and in the inlier [520560], north of the main outcrop, $\frac{1}{2}$ mile N. of Sevenoaks Station, Riverhead, chert is absent.

Fossils are locally common in the Hythe Beds and include the following. Brachiopoda: *Arenaciarcula fittoni* (Meyer), *Sellithyris sella* (J. Sowerby), *Sulcirhynchia hythensis* Owen, *Tamarella tamarindus* (J. de C. Sowerby).

Bivalvia: *Aptolinter aptiensis* (Pictet and Campiche), *Chlamys robinaldina* (d'Orbigny), *Entolium orbiculare* (J. Sowerby), *Exogyra latissima* (Lamarck), *Limatula tombeckiana* (d'Orbigny), *Linotrigonia* (*Oistotrigonia*) *ornata* (d'Orbigny), *Modiolus aequalis* J. Sowerby, *Plicatula carteroniana* d'Orbigny, *Pseudaphrodina ricordeana* (d'Orbigny), *Pseudolimea parallela* (J. Sowerby), *Septifer sublineatus* (d'Orbigny), *Thetironia minor* (J. de C. Sowerby), *Trigonia carinata* J. L. R. Agassiz.

Nautiloidea: *Anglonautilus undulatus* (J. Sowerby) *Cymatoceras spp.*

Ammonoidea: *Tropaeum spp., Cheloniceras spp., Dufrenoyia spp.*

Belemnoidea: *Neohibolites ewaldi* (von Strombeck).   H.G.D., R.C., C.J.W.

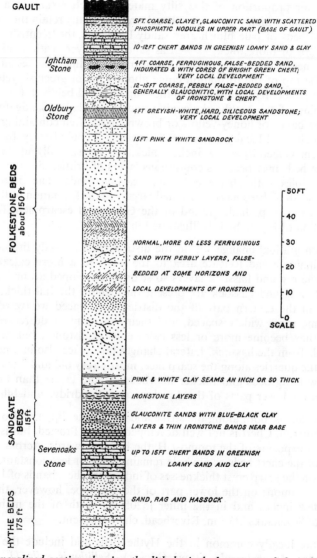

FIG. 3. *Generalized section showing the lithological characters of the subdivisions of the Lower Greensand (excluding the Atherfield Clay)*

(A 9146)

A. NORMAL FAULT, QUARRY HILL BRICKWORKS, TONBRIDGE

PLATE V

B. ALTERNATION OF RAG, CHERT AND HASSOCK IN HYTHE BEDS; NEAR OFFHAM

(A 10277)

Sandgate Beds

Throughout the district this formation, with a thickness varying from place to place between 2 and 20 ft, is characterized by a high glauconitic content and general clayey or loamy texture. In well sections it has been shown to be 2 to 12½ ft thick at Sundridge, 3 to 5 ft at Trottiscliffe and 11½ ft at Brasted, and again at Riverhead. The unweathered deposit has generally a bottle-green colour due to its high content of glauconite, which on weathering is oxidized to produce red, orange and yellow mottlings. The apparent clayey nature of the formation as a whole appears to be due to the softness of the glauconite, hence in weathered material a deep red colour frequently indicates a clayey texture presumably resulting from the breaking down of almost pure glauconite sand, while the paler yellow and orange colours accompany a higher content of siliceous sand.

Slight changes in lithological character are observable here and there. In several localities, especially south of Brasted and around Sevenoaks, thin limonitic seams occur in the bottom foot or so; while in the more northerly exposures around Riverhead the formation develops some true clay beds and, in places, e.g. east of St. Mary's Church, it is composed almost entirely of black or dark bluish green laminated clays, certain layers of which resemble fuller's earth, which may also occur as thin bands in the Sandgate Beds in other parts of their outcrop. There is a general tendency for coarse sand grains to increase in quantity downwards to form a pebbly clayey sand base.

Though the formation is thin it is traceable without difficulty across the district by virtue of its distinctive characters as compared with the more arenaceous formations above and below, although over wide stretches sections are rare. The main outcrop is narrow, seldom exceeding 150 yd, but it covers broader areas in the outliers and the salients that extend southwards from the main crop and mantle the dip-slope of the Lower Greensand escarpment. This is notably so between Brasted and Stone Street, where the formation appears to have contributed largely to the matrix of the Angular Chert Drift and to have imparted to it a red colour not only where the latter extends over the outliers but at higher levels on the dip-slope from which all traces of Sandgate Beds have now been removed. H.G.D., S.C.A.H.

Folkestone Beds

The outcrop of the Folkestone Beds extends across the district from Limpsfield in the west to Birling in the north-east. Generally occupying the lower part of the dip-slope of the Lower Greensand, it seldom exceeds ½ mile in width, but between Seal and Ightham, where a local facies of hard chert beds lies near the top of the formation, it widens to 1¾ miles. The thickness varies considerably and averages about 150 ft. The greatest known development in the district is in the west, where boreholes near Limpsfield of the East Surrey Water Company proved a thickness of 224 ft. Other thicknesses recorded from boreholes are 115 ft at the Sevenoaks Brickworks, Greatness, north of Sevenoaks, 179½ ft at Brasted and 106¾ ft at the Sevenoaks Waterworks Company's pumping station 1¼ miles N.E. of Seal. In the north-eastern corner of the district, at Wrotham Heath, the thickness is estimated to be 180 ft and at Trottiscliffe, as proved by boreholes, it is in the neighbourhood of 185 ft; ½ mile N.N.W. of Ryarsh about 173 ft were proved.

The Folkestone Beds of the present district have a somewhat varied lithology. They consist, in the main, of current-bedded sands, locally white or 'silver', but usually more or less ferruginous, originally deposited in very shallow water, and in part possibly of sand dune origin; they are generally of medium or coarse grade, occasionally pebbly, but in some layers fine-grained or even silty (Brown 1941, pp. 1–15; Casey 1946, pp. 43–7). Bands of ironstone or carstone up to 6 in or so thick and ramifying irregularly across the bedding are to be seen in most exposures (see Plate VIIB), and there are also variably sloping tabular bands of hard white sandstone. This type of deposit makes up the full thickness of the formation in the west, but around Brasted, Sundridge and Chipstead a fairly fine-grained, white or pale pink sandrock occurs in the upper part of the formation; its thickness is estimated to be at least 30 or 40 ft. Though generally soft and easily excavated it is sufficiently cohesive for quarries to assume the form of chambers or tunnels, and the rock is locally cemented into a hard sandstone, fragments of which, often of large size, occur in local drift deposits, and recall the sarsen stones of some Tertiary deposits.

The strong development of persistent hard chert beds between Seal and Ightham modifies the simple topographical form of the Lower Greensand dip slope by forming a subsidiary escarpment (Fig. 4). At the top of the formation in this area come 10 or 12 ft or more of loamy, glauconitic sand and clay with a series of chert beds from a few inches to a foot or so thick, repeating the lithology of the Sevenoaks Stone horizon at the top of the Hythe Beds (see Fig. 3). Cherts were noted at Westerham and are known to occur sporadically farther east at Brasted, Chipstead and Greatness. Between Greatness and Seal they are in the form of scattered nodules which have grown around fossil sponges in coarse ferruginous sand, but east of Seal they are traceable as continuous beds in glauconitic loams cropping out from below the Gault for a distance of about 2 miles, and were penetrated in wells at Sevenoaks Waterworks pumping station, 1 mile S.E. of Kemsing. As far to the north-east as Trottiscliffe several boreholes for the Mid Kent Water Company penetrated hard rock at the equivalent horizon, in sands just beneath the Gault. The southern limit of chert follows the crest of the subsidiary escarpment from Seal to Raspit Hill or Ightham Common, east of which beds of this type are not known to crop out. Immediately below the chert horizon a bed of coarse false-bedded sand, 3 or 4 ft thick, in places glauconitic and else-where limonitic, is cemented here and there into a sandstone with cores of bright green chert known as Ightham Stone (Drew *in* Topley 1875, p. 140). The extent of this rock type is difficult to trace. Its chief locality lies in the eastern parts of the ancient camp that surmounts Oldbury Hill, and south of the bridle path that crosses the camp from Styants Bottom to Ightham, but in the sunk roads leading up to the camp from the south and from the west true Ightham Stone is missing, though in the former cutting there is a green, but not cherty, sandstone at the same horizon. Owing to its mode of occurrence Ightham Stone is difficult to detect in the field and it may well be present at other localities.

Green chert fragments are common in the local accumulations of Angular Chert Drift on both Oldbury and Raspit hills, as well as in hillwash on the slopes, and it is probable that Ightham Stone *in situ* was once much more widespread than its present known position within the confines of Oldbury Camp would indicate. It is of interest that fragments of green chert comparable

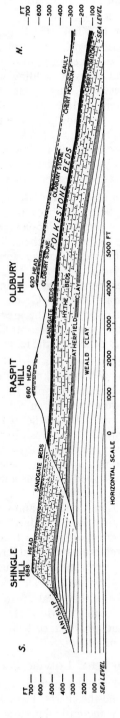

FIG. 4. *North–south sections across the Lower Greensand outcrop through Shingle Hill and Raspit and Oldbury Hills (see Fig. 1). The subsidiary escarpment at Oldbury Hill is capped by beds of chert at the top of the Folkestone Beds*

with Ightham Stone have been found in a small spread of river gravel in the Thames Valley near Cooling, 5 miles N. of Chatham.

Below the horizon of Ightham Stone come 12 to 15 ft of strata that are very variable from place to place. On the whole they consist of coarse pebbly sand, generally more or less glauconitic and with local ironstone and chert. Individual layers are lenticular and there is evidence both of channelling accompanied by steeply foreset false-bedding, and of even bedding under tranquil conditions in shallow basin-shaped depressions, suggesting formation under shallow coastal conditions with intermittent periods of partial emergence with shallow lagoons. Next in the downward succession is a very local hard, greyish white, siliceous sandstone or quartzite averaging 4 ft thick, often referred to as Oldbury Stone, the only known occurrence of which is on Oldbury Hill, where it crops out intermittently for over 1000 yd along the eastern side of the camp and for about 100 yd on the north-west corner, east of Broomsleigh.

The names 'Ightham Stone' and 'Oldbury Stone' are often used synonymously. The former, which can be traced back to the early part of the 19th century, was definitely applied by Drew (in Topley 1875, p. 140) and adopted by Bonney (1888, p. 297) to refer to the green chert, but there appears to be no precedent for confining the latter name to the greyish white quartzite. Sir Edward Harrison states (in litt.) that Benjamin Harrison employed the term Oldbury Stone for both varieties, but, when making special reference to the green chert, used the expression 'Oldbury Green Stone'. Prestwich (1889, p. 277) apparently following Harrison, calls the green chert 'Oldbury Stone', though he refers to Bonney. Throughout the present memoir the two names are kept distinct and used in the senses defined in the text.

Below the Oldbury Stone horizon there is a bed of pink and white mottled, fine to medium-grained sandrock, with a thickness of about 15 ft, to be seen in most exposures around Oldbury Hill and Raspit Hill. The well-known Palaeolithic rock shelters at Oldbury are excavated in this bed on the eastern side of the camp (Plate IXB). This sandrock may lie at the same horizon as that noted between Brasted and Chipstead, to the west, but it has not been observed between the latter place and Seal; it extends eastwards into the area around Borough Green, where a local inn built upon it is named the 'Rock Tavern'. The beds of the Seal–Ightham area described above account for about 50 ft of the top part of the Folkestone Beds, below which they are normal ferruginous sands though locally, near the bottom, thin pink and white pipeclay seams occur and, at the base, a few feet of fairly evenly bedded ironstone seams. Around Wrotham Heath the basal beds are more or less loamy. In the north-eastern part of the district the formation maintains its usual ferruginous sandy character throughout. This sandy character gives rise, over much of the outcrop of the Folkestone Beds, to sandy heathland (Plate IXA).

Small pebbles occur commonly in the basal 20 to 30 ft of the Folkestone Beds and also in the upper part, notably at Moor House and Westwood Farm, near Westerham. Their nature and provenance have been discussed by Wells and Gossling (1947, p. 195), among others.   H.G.D., S.C.A.H.

PETROGRAPHY OF THE LOWER GREENSAND

The Hythe Beds contain three distinct petrographic facies, the calcareous Kentish Rag, the sandy hassock and the cherts. These are linked by transitional types. The Kentish Rag is a grey glauconitic limestone (E 12438, 16909, 19361,

19366–7, 19852) contaminated with quartz grains. The rock is composed of the fragmental, partly recrystallized remains of echinoderms, brachiopods, foraminifera and bryozoa, in a matrix of clear crystalline calcite of grain size varying up to about 0·3 mm. The quartz grains are rounded or subangular, very ill-sorted, but not exceeding 0·75 mm in diameter. Small rolled pebbles of calcite mudstone occur, and fragments of chloritic sandstone, quartzite and chert may be present. The glauconite generally takes the form of pellet-shaped aggregates of sub-microscopic crystals, but it has also been noted as a filling of trabecular tissue in echinoderm fragments. Small pieces of collophane are common; fresh microcline was noted as a rare detrital constituent. Calcareous glauconitic sandstone, occurring for example in an old pit [45665192] 490 yd N.E. of Chartwell (E 16910) is a type transitional to the arenaceous facies.

The sands, compared with those of the Wealden Series, are often only roughly graded, carrying rounded quartz grains up to 0·5 mm diameter mixed with subangular grains averaging 0·1 mm. Glauconite is considerably more abundant than in the Wealden sands, but may not be present in sufficient quantity to colour the sands green. Consolidated sandstones occur in the hassock. A typical example (E 16916) collected at 44665280 in Tower Wood, 1 mile S.S.E. of Westerham, is composed of subangular quartz grains averaging 0·1 mm, with a few rounded grains up to 0·5 mm diameter. A light fraction, with specific gravity below 2·60, contains angular fragments of fresh microcline and albite (E 21650) with nontronitic material. A heavy fraction (E 21652, 3, 7) over 2·74 specific gravity, contains abundant pellets of glauconite with aggregate refractive index between 1·583 and 1·610; colourless almost uniaxial mica, and heavy mineral grains in the following percentage proportions (excluding glauconite and mica): zircon 35; magnetite and ilmenite 27; leucoxene 19; rutile 7; limonite 4; tourmaline 3; kyanite 3; staurolite 2; and rare or isolated grains of pyrite, monazite, sphene and malachite or chrysocolla. The zircon grains rarely exceed 0·05 mm, rounded forms being more abundant than euhedral; whereas both kyanite and staurolite grains reach 0·15 mm. Owing to the persistent presence of the latter two minerals, residues from the Lower Greensand may readily be distinguished from those of the Hastings Beds of the Wealden. The cement of the sandstone is mainly composed of opaline silica.

Intermediate petrographic types link the hassock sandstone with the cherts characteristic of the upper 12 ft of the Hythe Beds, all having a matrix composed of secondary silica in isotropic opaline form, or as chalcedony (E 16898–16908; 16911–15; 16917). The important role played by the spicules of monactinellid and tetractinellid sponges in the formation of these beds appears first to have been recognized by Hinde (1885, p. 403). The spicules are now mainly represented by hollow casts, but it seems probable that much if not all the silica forming the matrix of the cherty sandstones and cherts was derived from these siliceous organisms. The birefringent chalcedonic variety of silica has a maximum refractive index of 1·538 and includes fibrous and spherulitic material of the type classified by Cayeux (1916, p. 200) as "quartzite". The isotropic opaline variety frequently has a greenish tint; its refractive index commonly varies between 1·534 and 1·537, much above that of normal opal, and this may perhaps be due to hydrous iron silicate in solution. At some stage these cherty rocks must have been in the state of silica gels, with detrital quartz dispersed through them. Some of the quartz grains show embayed outlines suggesting that they have been partly replaced by secondary silica.

The Sandgate Beds include glauconitic sands in which the glauconite locally predominates over quartz. An example (E 21309) from the roadstone quarry at Dryhill (p. 71) is mainly composed of pellets of glauconite with aggregate refractive index between 1·605 and 1·615, some of which are partly converted into limonite. Quartz occurs in well-rounded and subangular grains averaging 0·2 mm but reaching 0·75 mm; fragments of quartzite, chert and mudstone are present and the rock is cemented by limonite. The same quarry shows a sandy limonite rock (E 21310) which possibly represents the ultimate stage of the conversion of glauconite to limonite. A glauconitic sand from a quarry [603566] 600 yd W. of Basted House, near Borough Green (E 18403, 21654–5), 5 ft above the base of the Sandgate Beds, was analysed with the following results:

Analysis of Glauconite Rock, Basted Quarry, Ightham

SiO_2 60·21; Al_2O_3 7·91; Fe_2O_3 17·69; FeO 0·35; MgO 1·27; CaO 0·52; Na_2O 0·10; K_2O 2·21; $H_2O > 105°$ 5·40; $H_2O < 105°$ 3·93; TiO_2 0·42; P_2O_5 0·11; MnO 0·04; CO_2 0·02; SO_3 0·01; FeS_2 0·03; Cr_2O_3 0·04; Li_2O Tr.; C 0·15; Total 100·41. Analyst: G. A. Sergeant 1939, Geological Survey Lab. No. 1062. Separation at 2·74 specific gravity gave a heavy crop amounting to 25% of the sample, composed mainly of glauconite pellets, some of them partly altered to limonite (E 21654). Very small amounts of the following minerals were also identified: ilmenite, leucoxene, rutile, zircon, tourmaline, kyanite, staurolite. Some of the zircon grains show marked zoning and segregation of dark inclusions to the centres of the grains, similar to those figured by Groves (1931, pl. VI) as characteristic of a Dartmoor provenance.

In the Folkestone Beds both poorly consolidated sands and secondary silica rocks again appear. An example of the former type, collected 250 yd S.W. of Combe Bank House (E 21306), is a friable white sandstone composed of well-rounded and subangular quartz grains averaging 0·1 mm with a few grains of mudstone and chert, and rare flakes of white mica. A heavy concentrate (E 21656), over 2·74 specific gravity, amounted to 0·45% by weight, and contained percentage proportions of species (based on a count of 750 grains) as follows: leucoxene 29, zircon 26, magnetite and ilmenite 19, tourmaline 8, rutile 7, kyanite 6, staurolite 2, limonite 1; with rare isolated grains of pyrite, anatase, brookite, hornblende and ?topaz. The suite is closely similar to that from the Hythe Beds (p. 63). This sandstone shows secondary quartz, in optical continuity with the quartz grains, forming a partial cement. Complete cementation by secondary quartz is displayed by the so-called Oldbury Stone (p. 62), which is a white quartzite composed of subangular quartz grains of 0·2 mm average diameter (E 21298). Small apatite crystals were noted as inclusions in some of the quartz grains in this rock.

The Ightham Stone, the petrography of which has previously been described by Bonney (1888, p. 297) is a green sandstone of glassy appearance, flecked with brown spots, which occurs as cores in a porous limonitic sandstone (E 21299). It is composed of well-rounded quartz grains averaging 0·4 mm, with a small proportion of grains of limonitic sandstone, limonite, shaly siltstone and chert. The matrix of the rocks consists of fibrous chalcedony with refractive indices $\alpha = 1·532, \gamma = 1·539$, the fibres being orientated perpendicular to the boundaries of the quartz grains. Cores of green-tinted isotropic material with refractive index 1·538–1·540, possibly an amorphous mixture of opal and iron silicate, occur in the matrix. In a rock closely resembling the Ightham

Stone, collected from a road cutting north-west of Styants Bottom (E 21303) the matrix is mainly isotropic, with refractive index ranging up to 1·550, but shreds of crystalline glauconite are present in it. The limonitic sandstone in which the Ightham Stone occurs differs from it only in that the matrix is partly made up of limonite, in some places as dense masses, in others as a dispersion of tiny fluffy particles through the green isotropic material. At about the same horizon as the Ightham Stone, on the west side of Oldbury Camp, a coarse pink porous sandstone (E 21302) contains abundant sponge spicules preserved in opaline material. These, together with quartz grains up to 2 mm diameter, and fragments of chert and sandstone, are enclosed in a chalcedonic matrix.

In Marriage's pit [520567], ½ mile E. of Riverhead (p. 82) a fault rock from the Folkestone Beds (E 21308) proves to be a brecciated sandstone in which quartz grains up to 0·5 mm diameter have been shattered and cemented with a mixture of limonite and chloritic or glauconitic material. Barite is sparingly present in this rock, but its rounded form suggests that it may be detrital.

K.C.D.

GAULT

The combined outcrops of the Gault and the Upper Greensand, the latter formation being absent eastward of Dunton Green, occupy the floor of the vale at the foot of the Chalk escarpment. Over much of its extent the vale lies at about 200 ft above Ordnance Datum, but west of Westerham and north of Oldbury Hill the base of the Gault crops out, at its highest point, at 400 ft O.D. In the area north of Ryarsh the Gault outcrop forms higher ground than the valley floor and, lying at about 200 ft O.D., borders the drift-filled valley cut into the Folkestone Beds by the stream near Leybourne Grange.

The Gault is essentially a dark blue or greyish homogeneous clay with a heavy, sticky texture, but the base is very sandy, with coarse pebbly grains, and much dark green glauconite and occasional thin sandy seams occur, such as the 'dark loamy silt', 14 or 15 in thick, recorded 88 ft down in the borehole at Trottiscliffe Rectory mentioned below. When weathered the clay frequently shows mottled orange and yellowish hues. Phosphatic nodules are abundant in the basal beds, often large and arranged in layers. A well-marked nodule band also occurs at the top of the Lower Gault. The clay is often noticeably micaceous and the lime content increases upwards near the junction with the Lower Chalk, the calcareous Gault being comparable with the Chalk Marl in composition and appearance, and indeed the two may easily be confused in the field. Glauconite is present throughout, usually only as comparatively scattered grains, except in the concentrated dark green clayey sand of the basal beds. Occasional pale limy seams are sometimes to be found at all horizons, and somewhat rarely the clay shows a shaly development.

The thickness of the Gault, as proved by boreholes, lies between 226 ft at Shoreham Place, in the Darent valley ¼ mile N. of the edge of the map area, and 323 ft ½ mile N. of Kemsing. At Titsey Court [40985495], just south-east of Titsey church, a borehole section shows 271 ft of Gault, while at Combebank Farm [478568], 1 mile S.W. of Chevening, a well-section indicates a total calculated thickness of about 250 ft. A well [52355940] ¼ mile W.N.W. of the church at Otford, starting somewhat below the top of the Gault, indicates an original thickness of about 320 ft, while the borehole at Trottiscliffe Rectory [644603], referred to above, commencing just below the top of the Gault, had not reached the base of the formation at 183 ft. The thickness of the Gault is thus seen to vary from place to place with a maximum development around Otford and Kemsing.

On palaeontological and lithological grounds the Gault may be divided into Lower Gault and Upper Gault, these two divisions corresponding approximately to the Middle and Upper Albian substages of international nomenclature. In this district the Lower Albian *mammillatum* Zone, which elsewhere may form part of the Lower Greensand, is more conveniently grouped with the Gault. The clays are capable of fine subdivision on the basis of the occurrence of different species of ammonites. In the following table of zones and subzones, modified from Spath (1942) and Casey (1961), horizons that have been recognized in the Sevenoaks area are indicated thus*.

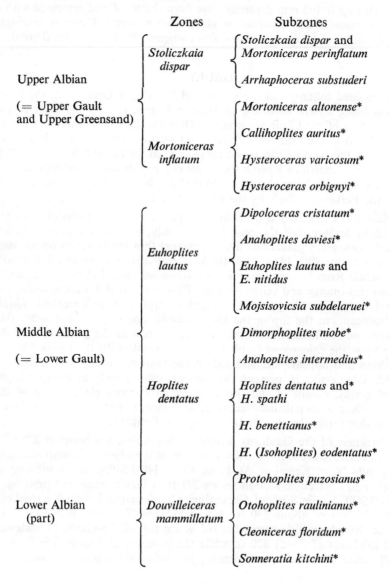

	Zones	Subzones
Upper Albian (= Upper Gault and Upper Greensand)	*Stoliczkaia dispar*	*Stoliczkaia dispar* and *Mortoniceras perinflatum*
		Arrhaphoceras substuderi
	Mortoniceras inflatum	*Mortoniceras altonense**
		*Callihoplites auritus**
		*Hysteroceras varicosum**
		*Hysteroceras orbignyi**
Middle Albian (= Lower Gault)	*Euhoplites lautus*	*Dipoloceras cristatum**
		*Anahoplites daviesi**
		Euhoplites lautus and *E. nitidus*
		*Mojsisovicsia subdelaruei**
	Hoplites dentatus	*Dimorphoplites niobe**
		*Anahoplites intermedius**
		Hoplites dentatus and* *H. spathi*
		*H. benettianus**
		*H. (Isohoplites) eodentatus**
Lower Albian (part)	*Douvilleiceras mammillatum*	*Protohoplites puzosianus**
		*Otohoplites raulinianus**
		*Cleoniceras floridum**
		*Sonneratia kitchini**

A. Basal Beds of the Gault, Borough Green

(A 7125)

PLATE VI

B. Quarry in Folkestone Beds, near Trottiscliffe

(A 7170)

The chief features of the succession in the Sevenoaks district as compared with that of the type section of the Gault at Folkestone, Kent, are (1) the expansion of the *dentatus* Zone, which attains a thickness of over 50 ft, more than twice that present at Folkestone, (2) the attenuation of the overlying *lautus* Zone and (3) the passage of the highest beds into a sandy facies (Upper Greensand). The reduced thickness of the *lautus* Zone is thought to be due to a shallowing of the sea or to changes in the submarine erosion level, resulting in the formation of phosphatic nodule beds (Spath 1942, p. 756; Owen 1963, p. 37). A concentration of such nodules, with a condensed fauna of *lautus* age, marks the top of the Lower Gault in the area of the Sevenoaks Sheet. The basal beds of the Gault (*mammillatum* Zone) show great variation from place to place and on palaeontological evidence include equivalents of the topmost 10 ft of the Folkestone Beds of the Lower Greensand as developed at Folkestone (Casey 1961, p. 544). There is evidence of a basin structure in *mammillatum* times with a centre in the area between Brasted and Seal and whose flank is exposed at Squerryes pit, Westerham (Casey 1961, pp. 543–6). The Upper Greensand has yielded no fossils diagnostic of horizon, though it is presumably of *dispar* age. S.C.A.H., R.C.

UPPER GREENSAND

The Upper Greensand consists of micaceous glauconitic malmstones and clayey sands. It crops out at the foot of the Chalk escarpment from westwards of Titsey to north-east of Westerham, narrowing from west to east in correspondence with a thinning of the beds from about 30 ft near the western margin of the area to not more than 3 or 4 ft at the watercress beds [45555563] 850 yd S.E. of Pilgrim House north of Westerham. Eastward the bed then dies out, to recur eastward again in lenticular fashion and finally to disappear in the Darent valley. A borehole at Jewels Wood (p. 149) in the north-west corner of the sheet, proved a thickness of 42 ft beneath Chalk.

The Upper Greensand is sharply folded or faulted near Limpsfield Lodge Farm, north of Limpsfield, from whence it falls gently eastward for a mile or so to rise about Pilgrims', formerly Pilgrims' Lodge, Farm eastward of Titsey. A small synclinal fold with a complementary anticline to the north lies between that farm and Clacket, ¼ mile E., but it flattens out in the next ¾ mile towards Gaysham and from that point the outcrop falls gently and the bed is almost free from subsidiary folding.

No pronounced feature marks the outcrop near the western edge of the area. Across Titsey Park, however, hard beds form surface features, and farther east minor folds have accentuated them. Eastwards the hard bands die out and features are slight, and usually the beds are obscured by a downwash of chalk and Clay-with-flints material from the Chalk escarpment. It is noticeable that this wash accumulates to the greatest depth over Chalk Marl and the base of the Chalk and over the clayey base of the Upper Greensand, the relatively harder beds of which lie close to the surface. S.B.

DETAILS

HYTHE BEDS

Beds of the upper horizons of the Hythe Beds in the vicinity of Limpsfield were noted in temporary sections which exposed grey and brown chert; sandstone with cores of grey chert; hard and soft, buff, glauconitic sandstone; light brown sand, frequently with glauconite; and brown, orange and red clays.

At Paines Hill the beds, as exposed in a roadside section [41135170] 300 yd N. of Bolthurst Farm, are folded into an anticline, and comprise soft glauconitic sandstones and sands, grey, buff and rusty brown in colour.

On the south-east of Limpsfield Common along the south side of the main road a series of quarries [425517, 430516] extends eastwards in which stone and sand have been worked at a variety of levels according to the demand for the different layers. The general section is as follows:

	Ft	in
Angular fragments of chert and sandstone in a clayey sand matrix	6	0
Chert	0	10
Soft sandstone (Hassock)	6	6
Hard sandstone (Walling Stone)	1	0
Sand	0	2
Yellow spicular sandstone (Square Block)	0	10
Cherty sandstone (Second Walling Stone)	0	8
Sand	0	4
Yellow sandstone with grey cherty cores (Rockery Stone) *Entolium orbiculare*	2	6
Cherty sandstone (Hard Stone)	1	0
Soft sandstone (Hassock)	3	0
Yellow spicular sandstone (Sandy Hassock) with fragments of silicified bored wood	5	0
Grey and buff sandstone. (Building Stone) *Anglonautilus undulatus*	1	0
Soft sandstone (Hassock)	1	0
Grey sandstone, yellow when weathered, speckled with glauconite (Building Stone)	1	0
Soft sandstone (Hassock)	3	0
Grey sandstone (Grey Building Stone)	1	0

The bracketed names are those applied to the beds by the quarrymen, according to whom many fossils have been found in the Square Block.

On the crest of the anticline at Moorhouse Bank (p. 12) is a small pit [42935325], 150 yd W.N.W. of Moor House, showing the base of the Sandgate Beds resting on chert. The underlying beds are hard grey sandstone and soft buff nodular sandstone locally passing into sand.

Brown chert was noted in Westerham, 350 yd N.E. of Squerryes Court and farther east on the northern bank of the Darent. It was also exposed immediately S. of Dunsdale [45555398], at Hosey Hill and in a pit [45675349] some 600 yd N.E. of Hosey Hill, where *Entolium orbiculare* was found.

East of Hosey Hill a line of old pits extends down the eastern side of a dry valley from a point [45445280] 500 yd S.E. of Chart's Edge to a point [45405316] 400 yd E.N.E. of the same house. These are mainly overgrown, but show 1 ft of sandy wash over 7 ft of buff, glauconitic sandstone in the south, and up to 10 ft of greenish buff sandstone with red spots in the north. At the last locality *Chlamys robinaldina*, *Entolium orbiculare*, *Exogyra latissima* and *Sellithyris sella* were collected. Below the 7 ft of sandstone, a good building stone has been quarried in underground quarries (Fig. 6). Numerous shallow pits in this neighbourhood have been worked into the east bank of the valley for sandstone until chert was exposed when the pits were successively abandoned.

North of Goodley Stock [44125273], the following section was seen:

	Ft	in
Sandy wash with chert fragments	2	0
Grey and brown mottled glauconitic sandy clay with angular cherts	1	4
Chert	2	6
Sand with grains of glauconite..	0	1
Chert	4	6
Sand with lenses of chert	0	8
Chert streaked with black	0	4
Buff clayey sand	0	1
Chert with black streaks	4	0
Chert, red clay and buff sand	1	0
Dark grey chert	0	10
Sand	0	2
Dark grey chert	0	8
Sand	0	3
Chert .. .. seen	1	0

On the high plateau at Crockhamhill Common, numerous exposures of chert associated with red clay occur in shallow pits opened in the overlying Head deposits. The chert is usually grey and dense, but occasionally includes a cream coloured, and rarely a white sugary variety. The chert does not appear to occur more than 40 ft below the highest point of the plateau, which is almost at the topmost part of the Hythe Beds. At a lower level on the escarpment sand and sandstone predominate.

About 700 yd N.N.E. of Chartwell the following section was seen in the highest beds of the formation [45705218]:

	Ft	in
Weathered grey sand with chips of chert	1	0
Pale buff clayey sand and reddish brown clay, mixed with small pieces of subangular bleached and sharp fresh chert up to 3 inches in length	2	6
Reddish brown clay with glauconite grains, and containing abundant lumps of fresh sharp chert up to 9 inches in length	1	2
Mauve-grey chert with yellow spots, locally bleached and sandy ..	0	7
Red loam, a clayey sand with grains of glauconite	1	6
Chert, locally bleached	0	6
Buff glauconitic sand	0	2
Reddish brown loam with glauconite, i.e. a ferruginous clayey sand	1	11
Almost colourless chert with pale yellow spots	1	2
Rough nodular buff bleached chert with green specks	2	6

In a small quarry [45665191] near the base of the Hythe Beds and 480 yd N.N.E. of Chartwell beds were exposed as follows:

	Ft	in
Brown clayey sand with lumps of sandstone	3	0
Soft, buff calcareous glauconitic sand and sandstone, weathering white with dark grey cores	4	0
Massive fine-grained calcareous sandstone	0	7
Lumpy buff sandstone, weathering white	1	5
Hard impure limestone with shells	0	8
Sandstone, locally with dark grey calcareous cores	4	0
Massive impure limestone (Kentish Rag)	2	0
Lumpy buff sandstone, weathering white	10	0

From the 8-in impure shelly limestone were collected *Aptolinter aptiensis*, *Pseudolimea parallela* and *Neohibolites sp.* A small pit 50 yd N.W. of the last locality [456520]

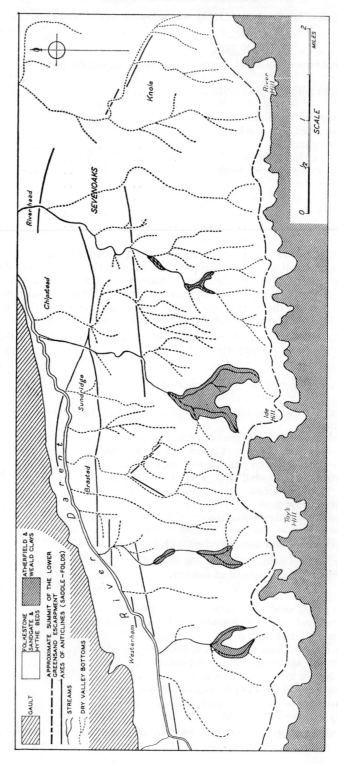

FIG. 5. *Sketch-map showing the relationship of the surface drainage of the Lower Greensand outcrop to saddle folds in the Westerham–Sevenoaks area*

exposed fossiliferous Kentish Rag with *Arenaciarcula fittoni, Limatula tombeckiana, Modiolus aequalis, Plicatula carteroniana, Pseudaphrodina ricordeana, Pseudolimea parallela* and *Septifer sublineatus.*

In the valleys near Colinette (formerly Brasted Place) Farm, the exposed strata, in places up to 20 ft, consist of alternations of 1-ft layers of sandy rag and 2-ft layers of sand, and in a pit [47945374] 800 yd S.E. of the farm, the rag and sand are overlain by about 10 ft of hassock with cherty layers.

On the lower part of the dip slope of the Lower Greensand and south of the Darent, there are numerous exposures of rag and hassock of the Hythe Beds in old quarries located on the anticlinal saddles that traverse the area from Westerham, as far east as Knole Park (p. 12). From The Quarry [47155487], 250 yd S. of the main road through Brasted Hill, to 300 yd E. of Sundridge church [48855498], a distance of little over a mile, beds exposed in a line of at least 15 pits along a saddle fold (see p. 12), some in one limb, some in the other, show dips ranging from 34° to 54° S. and from 34° to 56° N. The pit [47555494] 175 yd E.S.E. of Brasted Place exposed an arch of rag and hassock, with greenish grey glauconitic sand layers. On the northern limb of the fold rests a feather edge of red and green glauconite sand of the Sandgate Beds; chert layers are absent from the top of the Hythe formation in this locality. Eastwards the saddle fold was noted in the large roadstone quarry at Dryhill [49705516], where the eastern quarry-face, exposing rag and hassock, is over 70 ft high. The overlying Sandgate Beds on the northern limb rest on about 8 ft of chert bands in sand, which here form the top of the Hythe Beds.

The late H. Dighton Thomas (*in* Wright and Thomas 1947, p. 319, later modified by Casey 1961, p. 541) records that both the limestone and the sands are fossiliferous, but the latter yield the richer fauna, mainly as internal moulds, though the brachiopods, belemnites and the bivalve *Exogyra latissima* retain their calcite skeletons. The fossils of the sands include: *Serpula sp., Sulcirhynchia* cf. *hythensis* (common), *Tamarella* cf. *tamarindus, Cucullaea sp., Exogyra latissima* (abundant in all stages of growth from young to large adult individuals; also in the limestones), *Linotrigonia (Oistotrigonia) ornata* (abundant), *Pseudolimea parallela, Thetironia minor* (common), *Trapezium sp., Trigonia carinata* (rare), *Cymatoceras radiatum* (J. Sowerby) (common; also in the limestones), *Cheloniceras cornuelianum* (d'Orbigny), *Dufrenoyia sp., Tropaeum bowerbanki* J. de C. Sowerby (also in the limestones), *Neohibolites sp.* (abundant).

Other workings in rag and hassock were noted in pits about 300 yd N. of Dryhill [495553] and 700 yd N.E. of Sundridge church [49105538]. At 100 yd N. of Cold Arbor a narrow outcrop of Hythe Beds runs east–west with old pits at intervals. At two exposures 100 yd N.N.W. [51695532] and 200 yd N.E. [51845530] of Brittain's (formerly Britton's) Farm, in which the saddle folds have gentler dips than farther west, one shows loamy yellow sand with bands of glauconitic sandy rag up to 4 ft thick and lenses of chert up to 8 in thick with dips of 10° and 20° N. and the other a similar structure, the limbs dipping 28° S. and 23° N. The beds on the northern part of the face are turned upwards, forming a syncline, the northern limb of which dips 15° S. and abuts against a fault. To the north of this fault greenish yellow and red sandy clay of the Sandgate Beds rests upon thin layers of cherty doggers in sand dipping 30° N. This fold was described by Fitton (1836, pp. 133–6).

Excavations for the Sevenoaks By-pass in the Gracious Lane vicinity exposed deep sections in the Hythe Beds. A typical section beneath the Head is that where the bridge carrying the White House Road crosses the By-pass [52055245]:

Ft

Hard silicified cherty sandstone, top 4 ft broken and infilled with
   reddish brown sandy clay   5–8
Greyish green, brown in places, glauconitic clayey sand with thin
   irregular silicified bands   10–15

Ft

Wedge of sandstone with sponge spicules, irregularly glauconitic, friable and weathering white where unstained. Some planty debris. Stained with manganese and in places cindery and completely impregnated with manganese. Thinning and splitting to the south   0–8

Greenish grey, glauconitic sand with thin sandstone bands..   .. seen 12

A north-east–south-west trending solution fissure crossed the By-pass at this place. It averaged 1 to 1½ ft wide reaching a maximum of 3 ft wide and was proved to continue to at least 120 ft below ground level. The upper part of the fissure was infilled with Head composed of reddish brown sand with angular chert and sandstone fragments. The fissure may have opened originally as a result of cambering.

Farther north along the By-pass a cutting [51705292] exposed, beneath 6 ft of Head, 2 to 3 ft of bluish grey cherty sandstone overlying 1 to 2 ft of greyish green, glauconitic sandstone. The section just south of the Dibden Lane Bridge [51515370] revealed 2 ft of greenish glauconitic sand overlying 2 ft of purplish grey hard cherty sandstone with 4 ft of brownish grey sand and sandstone at the base of the section.

The railway cutting between Sevenoaks Station and the tunnel through the Lower Greensand escarpment, about 800 yd long and with a maximum depth of about 100 ft near the northern tunnel entrance, exposed many bands of rag and hassock with a variable gentle dip, mainly northwards.

An inlier of Hythe Beds, some 750 yd long and 70 yd broad, extending from 550 yd E. of St. Mary's Church, Riverhead, to just W. [52665597] of Linden Chase, in the St. John's district of Sevenoaks, is brought to surface by a saddle fold. Exposures of rag showing the arch are to be seen in the banks of a spring-fed pond [52145607] 700 yd E. of St. Mary's Church, Riverhead; also 70 yd E. of this point [52195606] and another around the tennis courts of a garden [52555600] east of Bradbourne Park Road. In the last, about 18 ft of greyish white, glauconitic sand with doggers of hassock and beds of massive rag up to 2 ft thick show a dip of 30° N. and are overlain to the north by red, green and yellow clayey sand of the Sandgate Beds, with cherty nodules at the base. The chert horizon of the Hythe Beds is absent here.

In the pit [53535520] at Knole Park (see p. 13) near the Lodge, 800 yd W. of Blackhall Farm, is a 15-ft face of grey glauconitic sand with massive beds of rag and layers of coarse hassock, some of which are brecciated, the beds being folded into a gentle syncline. At 400 yd S.E. [53815506] of this a complementary anticline lying to the north-east of the synclinal axis is to be seen, with dips of 7° S.W. and 9° N.E.; 600 yd farther S.E. are several old pits [54135469] showing rag, with dips up to 38° N.E. and 67° S.W. (see p. 13).

Other rag pits in this area are mostly overgrown; one at the road junction [53905595] 800 yd N.N.W. of Blackhall exposed in 1932:

Ft

Buff loamy sand with fragments of ironstone, chert and rag, pale brown at base (drift)   2

Clayey sand, coffee-brown coloured probably due to oxidation of glauconite   2

Greenish grey, glauconitic sand   3

Seam of sand with irregularly distributed chert nodules   1

Greenish grey glauconitic sand with scattered cherty nodules   ..   2

Alternating layers of large rag doggers, glauconitic sand and hassock, the last generally coarse-grained and with blebs of glauconite up to ¼-in diameter   ..   ..   ..   ·.   ..   30

The dip in this area is 2°.

At 400 yd [49755420], 1000 yd [50305428] and 1200 yd E. [50455430] of Manor Farm, south of Sundridge, small excavations along a saddle fold (p. 12) show massive rag with dips up to 23° S. and 51° N.; other pits along the same fold are at 1000 yd W. [51165439], 600 yd W. 5° N. [516544] and 350 yd N.E. [52275461] of Kippington House (formerly Bishop's Court), the last being a private rock garden, just west of the Sevenoaks railway cutting. On the east side of the southern end of the small inlier of Atherfield Clay [460535] west of Pipers Green, south-west of Brasted, a pit [46135346] just above the junction showed three 1½-ft to 2-ft bands of massive rag, separated by sand layers. Their dip of 10° S.W. may be due to slipping, for at the entrance to the pit are three step faults, 3 to 4 ft apart, each of which lowers the rag beds about a foot. Around the large Atherfield Clay inlier at Norman Street [490530] are several pits; rag and sand exposed along the east side invariably dip east, the inclination ranging from 5° to 17°.

The chert beds which occur near the top of the Hythe Beds are exposed here and there in the bottoms of the gravel workings in the Head drift. Near the summit of Toy's Hill [469517] 400 yd S.S.W. of the Fox and Hounds, 8 ft of red loamy sand with massive chert layers occur below 5 ft of drift. In an extensive excavation 600 yd S.W. [50405235] of Apps Hollow (formerly Apps Bottom) drift, about 6 ft thick, is underlain by a 2-ft layer of slabby chert, resting upon 2 ft of greenish yellow or brown glauconitic loam, below which are two 2-ft chert bands separated by a foot of sand. At 400 yd W. 7° S. of Shaw Well, a cutting for the drive to a private house [51755206] exposed, in 1932, 2 ft of angular chert rubble on 12 ft of massive chert layers up to about a foot thick in loamy sand, overlying 15 ft of yellow sand with occasional sandy doggers; another exposure [51655240] in the valley 550 yd N.W. of Shaw Well showed, below 5 ft of drift, two 4-ft bands of compact bedded chert, separated by 2½ ft of clayey glauconitic sand, red at the top and greenish to yellowish brown below. The chert contains sponge spicules and has encrustations of botryoidal opal on the faces of joints and cracks. A pit [54355268] on the scarp-crest, 700 yd E. of the main road at Riverhill, exposed in 1932 up to 10 ft of drift on about 10 ft of beds of massive chert with seams of loamy, greenish grey, glauconitic sand and whitish wisps of clay. The beds are slightly flexured but have a general dip of 1° N. Another exposure on the scarp-crest, 800 yd farther E.N.E. [55005287], showed similar beds dipping 5° to 8° N.

Along the scarp face of the Lower Greensand are numerous rag quarries, most of them long since abandoned, and sections in road banks. These usually show a certain amount of disturbance due to slipping and cambering of the Hythe Beds over the underlying Atherfield and Weald Clays. In the sunk track [497518] leading south from Goathurst Common to Yorkhill, the strata dip 4° N. near the crest, are horizontal about 150 yd S. of the east–west road along the crest, and farther south dip 18° S. Near the foot of the scarp-face and the base of the Hythe Beds there is greyish white sand with seams of sandstone. An old quarry [51275160] west of the road up the escarpment at Bayley's Hill is much obscured by a scree of greyish white loamy sand but shows massive beds of rag up to 2 ft thick, some of which appear to have a low lime content and to be a greyish white glauconitic sandstone; layers of sand with nodular hassock occur between them. A roadside section at the upper bend of the road up the hill [51535185] exposed weathered rag, pale grey and yellowish in colour, with ferruginous concretions. The blocks of rag are tilted slightly in all directions. At Hubbard's Hill, 1 mile E., rag is exposed in the roadside near the top of the escarpment, and also in old pits [52605197 and 52665197], just west of the road, near the base of the formation. From there to River Hill the scarp face is densely wooded and exposures rare; the hummocky ground surface here is indicative of slipping (see p. 143). At the top of River Hill [53875236], near the turning east to Fawke Common, chert beds dip 3° N. and rag is exposed below, east of the sharp bend in the road, where a quarry [54175224] shows dips of 6° N. on the north face and 9° S. on the south face. Much of the lower slopes of the scarp-face of the salient of Hythe Beds on which Riverhill House stands is hummocky and, according to local report, the ground is unstable, slow displacement of gate and fence posts having been observed. Several exposures in the scarp-face north-by-west of Under River show dips ranging from 3° to 18° N. and in an old quarry [55655287], west of the road up Carter's Hill, there is much disordered, weathered, greyish white, glauconitic sandstone with a tufaceous encrustation on exposed faces. In the roadside bank on Carter's Hill there are dips between 3° and 25° N. suggestive of slipping; near the road-bend about halfway up the hill, massive rag is exposed, overlain by 25 ft of grey, buff, loamy sand, with some rag doggers and a few cherty nodules near the top. Slight 'stepping' of the massive rag may be due to hill-creep. A superficial accumulation of about 6 ft of sand with rounded rag boulders overlies the solid in one place. East of Carter's Hill there are several large abandoned quarries on the scarp-face, known as White Rocks [56105305], showing up to 35 ft of rag, nodular near the top but massive below, with a small proportion of interbedded sand and hassock.

In the area of Hythe Beds between Godden Green and Stone Street and as far north as The Grove, south-east of Seal, many old excavations, all at high horizons in the formation, show rag and sand, and occasionally the chert bands, occurring near the top of the formation, the latter particularly just west of Bitchet Green where the thickness of the beds of the chert facies is 12 ft. *Exogyra latissima* was found 70 yd N.W. of Wilmot Cottages [57345320]. South of Wilmot (or Shingle) Hill there is a mass of landslipped Hythe Beds on the lower part of the scarp face [570527], but the outlying patch at South Seers Shaw [585529], 600 yd S. of Ightham Mote appears to be *in situ*.

Many old rag pits occur over the wide expanse of Hythe Beds between Ightham and Borough Green, all more or less overgrown and obscured. In 1936, a temporary roadside excavation [59005623] 200 yd E. of Oldbury Place, south-west of Ightham, showed a 2-ft rag bed, overlain by 4 ft of greenish yellow loamy sand with layers of chert doggers, and that by 6 ft of Sandgate Beds. In 1900, chert below tenacious clay of the Sandgate Beds was recorded in a temporary section in Ightham on the south side of the road to Borough Green and east of the road to Ivy Hatch (Harrison 1928, p. 15). The only excavations of note in this area are those about ½ mile E. of Ightham, of which the more westerly quarry [602565], in 1950, exposed up to 70 ft of Hythe Beds, worked in benches. The vertical face on the top bench exposed the Sandgate Beds, resting on the chert horizon of the Hythe Beds here about 3 ft thick. In a gentle synclinal depression affecting the top 20 ft of beds only near the centre of the face the base of the overlying Folkestone Beds rested on the Sandgate. The second bench showed 10 ft of rag and hassock, the former in layers of dogger-like masses up to 2 ft thick but averaging 6 in. The syncline brought the chert layers of the top of the formation into this face. The third bench showed 10 ft of massive rag beds with partings of hassock but not so much sand; in this face the syncline as seen on the top two benches had disappeared. The bottom three benches, each 10 to 12 ft high, were all in massive rag and hassock with very thin sand partings. Crushed *Tropaeum bowerbanki* and *Cheloniceras* cf. *cornuelianum* occur in these lowest hassock beds (Casey 1961, p. 540). False bedding was evident in some layers. The beds are almost horizontal but show a slight dip north-westwards. The cause of the depression in the upper part of the above section became evident when enlargement of the workings revealed strong fissures beneath two such down-sagging areas (Casey 1951, p. xxiv).

The large quarry [606567] 400 yd E. of the above shows 6 ft of superficial sandy material with rubble of chert and other local rock, resting on 3 to 5 ft of chert beds *in situ*, and these upon 40 to 60 ft of alternating 1-ft to 2-ft bands of rag and 2-ft to 4-ft layers of hassock. The strata are crossed by several vertical fissures, trending south-west, up to 4 or 5 ft wide in places, in part filled with sand, sandy loam resembling brickearth, and rock fragments. Above the filling, the fissures stand open in the rag and hassock, but in the chert horizon near the surface the walls come together, and the fissures are closed. In places the walls are coated with tufa, and stalactites and stalagmites of carbonate of lime occur, the latter growing on a tufa-covered surface of the filling, which is itself cemented to a depth of several feet; elsewhere the infilling sand and rubble is loose. The fissures are noteworthy on account of the Pleistocene and Recent fossil remains discovered in them (see p. 139). In the area south of Ightham and Borough Green numerous gentle depressions, generally oval in form, may in some cases be ancient rag pits that have been partly filled and ploughed over, but others may be due to subsidence on lines of fissures similar to those referred to above. The largest of these occur east of Ightham Park (or Ightham Warren) where, at Chalklins Shaw [59905615], a depression about 50 yd wide extends for 800 yd S.S.W. from a point [59985638] 500 yd E. of the road-turning east from the Ightham–Shipbourne road. Old excavations [59895615] in the east side of this showed 8 ft of sand with thin layers of rag with cherty cores. At 250 yd E. of this depression and parallel with it, is a second [60085600], 500 yd long, and several others in the vicinity are up to 100 yd long. Much smaller depressions occur in the area south of Claygate Cross and Crouch.

East of the Shode valley the Hythe Beds outcrop, which there attains its greatest width within the district, is extensively mantled by Angular Chert Drift, and good exposures are rare. Small exposures on the scarp-face usually exhibit disturbances of the strata due to slight slipping, but on the dip-slope the beds have a slight northerly dip, or are horizontal as in the pit [67155640] 200 yd N.N.W. of King Hill Hostel, formerly Malling Union Workhouse, where there are 6 ft of sand with 1½-in to 6-in layers of chert, resting on and in places piped into rag and hassock, seen to a depth of 6 ft, one 2-ft rag layer lying immediately below the chert.   H.G.D., S.C.A.H.

At Comp Farm, formerly Comp [64555716], about 1 mile W. of Offham, 9 ft of alternating rag and sandy beds showed a northerly dip of 4° or 5°, while 150 yd S. the beds in an old quarry [64555700] were seen to dip 4° N.E. Here 14 ft of Hythe Beds were seen to consist of ragstone with chert and in places glauconitic patches, alternating with beds of coarse glauconitic sandy marl; the individual beds average 1 to 2 ft in thickness. About 600 yd W.S.W. of Aldon horizontal chert beds near the top of the formation were observed [64405785]. Cherty limestone and sandy beds, each about 2 ft thick, were seen to a depth of 11 ft in a small pit [65095721] 600 yd E. of Comp Farm.

The large quarries just west of Offham showed in 1939 between 70 and 80 ft of Hythe Beds (Brown 1941, p. 13). The individual beds vary from 6 in to 3 ft in thickness and consist of variably glauconitic and sandy rag alternating with coarse glauconitic calcareous hassock. Beds of chert, dark bluish or black where unaltered, occur most frequently in the higher levels. They average 6 inches in thickness and appear commonly as siliceous or partly siliceous developments of the rag. The Hythe Beds lie horizontally (Plate VB) or with a slight north-easterly dip. In recent years more extensive working of the lower beds has exposed variable current-bedded strata, showing discordances and wedge formations. Some beds, however, are relatively persistent, for example the 'granny lane', which is a gritty ragstone, somewhat phosphatic at the base, averaging 2 ft in thickness and lying about 5 ft above the quarry floor. Though this bed dies out eastward, another ragstone bed, pale blue and very compact, which comes in beneath phosphatic hassock a little below, may persist into the Maidstone district. Fossils include common *Exogyra latissima* and the ammonites *Tropaeum benstedi* Casey and *Epicheloniceras sp.* from near the base (Casey 1961, p. 540). Drift which occupies solution pipes and hollows is described on p. 127.

Just north of the main road and about ½ mile E.S.E. of Addington church an old quarry [66155872] showed 16 ft of alternating rag and hassock in rather irregular beds each 6 in to 2 ft thick, and in lenticles. The hassock is here a glauconitic sandy marl compacted in patches into a soft calcareous sandstone. Nearest the road a northerly dip of about 6° indicates the proximity of a small anticlinal saddle-fold to the south (see p. 14) but northwards the beds gently flatten and become horizontal.

Along the line of the more extensive saddle fold (p. 14) between here and Offham
12 ft of sharply contorted rag and hassock with broken up chert bands [66055820]
were seen 150 yd N.N.E. of Offham church. Near the eastern end of the same line
10 ft of beds consisting of rag and calcareous sand or sandstone, with chert seams,

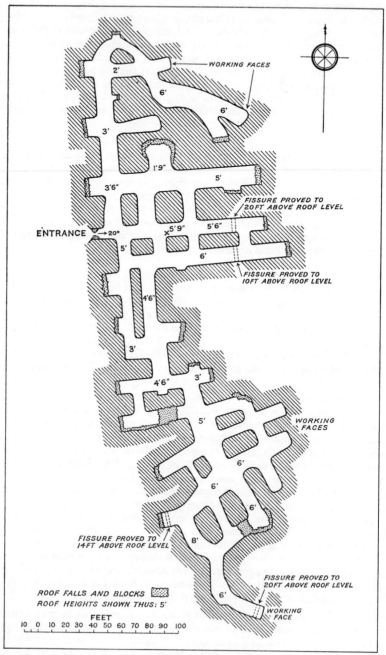

FIG. 6. *Plan of abandoned underground quarries in Hythe Beds at Hosey Common,
Westerham, Kent. From a survey made by Messrs. S. L. Birchby, G. A. Peet
and R. Rumbold, 1939 and 1947*

were seen in an overgrown pit [66905777] in a dell 200 yd N.N.E. of Fartherwell. The softer beds are irregularly pinched out owing to the sharp folding but the beds have also been disturbed somewhat by solution.

A degraded section [661571] 250 yd S.E. of the inn at Offham showed 6 ft of horizontal beds of rag and hassock and by the roadside ½ mile E. [670572], where lenticular developments and small scale current-bedding were noted, similar beds dip S.S.E. about 20°. Two roadside exposures [67685714] ¼ mile S.W. of West Malling church together showed 15 ft of alternating rag and hassock beds with chert; dips are variable up to 6° in this area.

North of West Malling and 200 yd S.S.E. of the inn (The Wheatsheaf) an overgrown pit [67825833] on the south side of the Maidstone Road exposed 9 ft of rag and hassock beds from 1 to 2 ft thick, with some thinner seams and lenticles, all dipping north 2°, and 200 yd E.N.E. [67905855] of The Wheatsheaf on the north side of the road, beneath the Sandgate Beds (see p. 79), 25½ ft of Hythe Beds were well exposed (Brown 1937, p. 397). At the top is a 6-in clayey bed seamed with thin calcareous sandstone and below this the beds consist of the usual alternating hard rag and soft glauconitic calcareous hassock. Though generally the individual beds are 1½ to 3 ft thick, the sandy beds are in places pinched out between limestones which may coalesce to form one bed perhaps 5 ft or so in thickness. Some of the sandstone is fairly hard, and chert layers, especially in the limestone, are developed in the lowest 10 ft of the section. The dip is north, about 3°. Fossils are common and include species of *Gervillella*, *Exogyra* and *Trigonia*. About 300 yd E.S.E. of this quarry 9 ft of horizontal rag and hassock beds were seen in a bank [68225840] beside the path across the stream from West Malling.

The large outlier south-east of Mereworth is much disturbed, most exposures showing dips in various directions (see, e.g., Brown and Himus 1938, p. 55). The disused quarry [67455275] about ¼ mile S.S.W. of Pizien Well exposed very shattered rag and hassock, in which a sequence of beds generally similar to those in the Maidstone district can be made out. Dips up to 40° show no consistent direction and it is clear that the beds have foundered as a result of strong cambering, a feature which was subsequently noted on the eastern border of the outlier, where it is bounded by the Medway valley. Irregular pockets of red loam which cap the Hythe Beds in this quarry are typical of a deposit which covers unevenly the higher parts of the outlier. Though it resembles brickearth it was probably derived from the remains of Sandgate Beds preserved in depressions, pipes and fissures in the rag and hassock beds. Two patches of slipped Hythe Beds material occur on the southern slopes of the outlier.

In the Hythe Beds several water boreholes northward of the outcrop have shown rag and hassock type of lithology to persist, with variations. A borehole ½ mile S. of Trottiscliffe proved 55 ft of rag and hassock beds which become silty and argillaceous and contain less stone towards the bottom. A little chert is present. The more hassocky and phosphatic cores yielded a large fauna including ammonites (Casey 1961, p. 540). At Sundridge glauconitic sands with sandstone and chert occupy the top 13 ft of the total 96 ft of Hythe Beds, and the basal 12 ft are calcareous sandy clays. Boreholes at Brasted have proved Hythe Beds, with predominant ragstone, to 122 ft without reaching Atherfield Clay. Sand, sandstone and chert occur in the top 20 ft and clays with pebbly sand in the lowest 30 ft. Occasional thin clayey beds intercalated with ragstone have been proved in various boreholes.   S.C.A.H.

SANDGATE BEDS

A Oxted Station, immediately beyond the western margin of the sheet and ¾ mile W.S.W. of Limpsfield church, a well passed through 30½ ft of sandy clay which may or may not all represent the Sandgate Beds, but boreholes [42505415] 1¼ miles E.N.E. of Limpsfield proved a maximum of 6½ ft (Whitaker 1912, pp. 205, 241).

On a roadside bank [40235196] 540 yd E.N.E. of the southern end of the railway tunnel near Limpsfield the Sandgate Beds comprise glauconitic sandy clay overlying a mixture of sand, very rich in glauconite, and orange-red clay. This rests on chert of the Hythe Beds.

Across Limpsfield Common glauconitic sand, orange clayey sand, mottled grey and red loamy clay and orange clay with glauconite occur. The brickearth pit on Limpsfield Common to which Topley (1875, p. 194) refers is in re-sorted Sandgate Beds.

To the north and south of the sharp anticline [430532] north of Moor House (p. 12) that has brought Hythe Beds to the surface, exposures of orange glauconitic clayey sand and sandy clay were noted. Similar material was seen on the northern limb of the fold where it crosses the county boundary [43295336]; in an old pit [43405333] on the crest of the fold 140 yd E. of the boundary, and again in a temporary section [444538] on the east side of the High Street at Westerham 500 yd S.W. of the church.

In an old pit [44255314] 340 yd S.S.E. of Squerryes Court orange sand with a thin bed of grey sandy clay overlies chert similar to that which occurs at the top of the Hythe Beds.

A feather-edge of Sandgate Beds, consisting of red and green glauconite sand, with many ironstone seams each about ¼ in thick at intervals of one or two inches, rests on rag and hassock of the Hythe Beds in the pit [47765496] 175 yd S.E. of Brasted Place, and the formation is also exposed on the north side of the large roadstone quarry [49855523] at Dryhill, 1 mile E.S.E. of Sundridge, where red and green mottled clayey sand rests on 3 ft of bottle-green glauconite sand and that upon 2 ft of red glauconite sand with thin ironstone seams. These beds lie on the northern limb of the saddle fold exposed in the quarry and are much contorted, but have a general northerly dip. The outliers on the dip slope of the Lower Greensand escarpment south of Brasted and Sundridge are not well exposed, but give rise to a reddish clayey soil which contrasts with the pale greyish brown and lighter sandy soil of the surrounding Hythe Beds. In shallow road cuttings south of Kippington House (formerly Bishop's Court, 52205435), up to 8 ft of red, green and yellow mottled, glauconitic, sandy clay with thin ironstone layers were noted.

Two sections along the line of the Sevenoaks By-pass exposed the full thickness of the Sandgate Beds. The first [50985455], on the western side of the new road ¼ mile S. of Salters Heath, showed beneath the Folkestone Beds 5 ft of false-bedded greyish brown sand with clay intercalations and bands of reddish iron-stained sand, resting on 7 ft of reddish brown clayey sand, strongly false bedded and with irregular iron-staining. The junction with the Hythe Beds was irregular. In the second section [50655500], 300 yd E.N.E. of Salters Heath, the Sandgate Beds were reduced to 5 ft of reddish brown false-bedded clayey sand.

The railway cutting at Riverhead is largely in Folkestone Beds, but 250 yd S.W. of St. Mary's Church, and 40 yd S. of a footbridge over the cutting, the top of the Sandgate Beds, brought up by the saddle fold through the Bradbourne Hall Estate to the east, is exposed for about 10 ft along the bottom of the cutting [517560]. The strata consist of toffee-brown to buff sandy clay, resting on stiff, brown and yellow clay with 4-in to 6-in seams of bluish grey shaly clay, olive-green in colour when fractured across the lamination and resembling fuller's earth in texture. About 100 yd E. of the cutting a pit [51755603] exposed the full thickness of the Sandgate Beds, consisting of 4 to 5 ft of black clay, weathering red and orange near surface, on 2 ft of clay like fuller's earth, greyish but stained yellow above and dark blue below. In another pit [51925606] a little farther east the formation consists of greenish grey compact sandy clay, the bottom 6 in to 1 ft of which contains paper-thin ironstone layers.

East of Bat and Ball railway station, Sevenoaks, an inlier of Sandgate Beds gives rise to swampy ground and to springs from the base of the Folkestone Beds which feed the pond of the corn mill south of Greatness. Ponds in the Wildernesse and in Knole Park are also on this formation.

A road cutting [53565310] over ½ mile S. of Sevenoaks church is in a red and orange coloured clay with glauconite. The clay here appears to have resulted from the almost complete oxidation of glauconite sand, and also in the old brickpit [53335314] a little farther south where 6 to 8 ft of red, orange and yellow sandy clay with glauconite and a little mica were exposed beneath 3 to 5 ft of local drift. The redder parts are more clayey than the yellow, which are more sandy, and small pellets of unaltered bottle-green glauconite sand are present.

Around Fawke Common, Godden Green, Bitchet Green and Ivy Hatch indications of red, green and orange clays appear over most of the outcrop where the soil consists of reddish loam. In 1936 a temporary roadside excavation [59005623], 200 yd S.E. of Oldbury Place, south-west of Ightham, showed 6 ft of green, yellow and red mottled glauconite sand resting on 4 ft of greenish yellow sand with chert nodules, forming the top of the Hythe Beds.

In the Basted quarry [603566], about ½ mile E. of Ightham, the full thickness, 16 ft, of Sandgate Beds was exposed in 1936, as follows:

		Ft
Folkestone Beds	Orange-coloured coarse pebbly sand with 3-in layers of ironstone near the base: seen in a synclinal depression in the strata (see p. 74) up to	6
Sandgate Beds	Greyish green and orange brown, evenly bedded clayey sand	6
	Similar to the above but paler coloured, with sand of finer texture and more plastic	5
	Bottle-green glauconite sand, stained black in places (For an analysis of sand see p. 64)	3
	Dark ink-coloured laminated clay with dark greenish grey sand streaks, irony concretions here and there and ochreous staining at the base	2
Hythe Beds	Greenish grey and brown loamy sand with bands of chert doggers	3
	Rag and hassock	–

Eastwards from this point there are only occasional small exposures in road banks, ditches etc., across the outcrop, which show red, orange and green mottled glauconitic sandy clays.   H.G.D., S.C.A.H.

From Borough Green to Addington and beyond, the narrow outcrop of the Sandgate Beds is marked by the presence of characteristic orange-weathered strongly glauconitic loam, with the glauconite frequently much decayed. From the Addington district to West Malling the Sandgate Beds have a very narrow continuous outcrop, largely drift-covered, on the north side of the stream flowing to Leybourne (just east of the map area) and as a series of outliers south of the stream.

Along the main outcrop the only large exposure seen was the section above the Hythe Beds in the quarry [67905855] 200 yd E.N.E. of the Wheatsheaf Inn on the Maidstone road north of West Malling (p. 77). This showed Head deposits composed of from 1 to 5 ft of sandy drift with flints and ironstone on Sandgate Beds comprising 3 ft of brown glauconitic sand with sporadic thin ironstone, resting on 6 ft of highly glauconitic greenish brown sandy loam, very fine and clayey in the lower half, on

1 ft of black and brown glauconitic fissile clay with sand grains and sandy patches. This rested on Hythe Beds seen for 25½ ft.

The junction with the Hythe Beds is sharply defined but, as described by Brown (1937, p. 397), it is irregular, forming a series of solution pockets at the base of which there are occasional small pebbles. The junction with the Folkestone Beds must be only a little up the slope above the top of the quarry beneath the sandy drift mantling the hillside.

Glauconitic loam is not well exposed in the three small outliers, two of which are capped by Folkestone Beds, between Addington and Offham, but a thickness of 10 ft was seen [66075824] 200 yd N.N.E. of Offham church, on the northern side of the sharp anticline (p. 14) which there separates the Sandgate Beds outcrop into two parts. In two larger spreads south-east of Addington and at West Malling, also capped by Folkestone Beds, glauconitic clayey loams are but poorly exposed, but Sandgate Beds of an estimated thickness of 8 ft crop out in the railway cutting at West Malling; nowhere do the beds appear to thin out completely (cf. Brown 1937, p. 396). In the outlier [68235723] south-east of West Malling yellowish brown glauconitic sandy loam with ironstone was seen in the road bank about ¼ mile S.E. of West Malling church.   S.C.A.H.

FOLKESTONE BEDS

Immediately south-east of the Coach and Horses Inn, Limpsfield and 500 yd S.E. of the church a section [408528] showed 18 ft of coarse current-bedded ferruginous sand veined with ironstone, with a pebbly bed at the base, overlying 3 ft of fine even-grained pink sand, resting on medium-grained, current-bedded yellow sand with coarse seams and pink layers and large pink spots, seen for 28 ft.

About 400 yd N.W. of Moor House, near the inn, a pit [42805345] showed 1 ft of chocolate-coloured soil with flints and pebbles, on light chocolate brickearth, with few stones, up to 5 ft thick lying on 20 ft of medium to fine-grained yellow sand and ironstone. Iron stains form concentric patterns; there are no pebble beds, but locally current-bedding on a fine scale is present. This bed rests on coarse sand. The medium to fine-grained sand rests on an irregular surface of coarse sand which has a general dip at a small angle to the north.

Another pit [428538], 150 yd S. of Westwood Farm, showed current-bedded coarse, medium and fine-grained yellow sands, the foresets dipping south, with undulating irony partings. The beds are generally pebbly, but less so below a depth of 6 ft from the surface than at the top. Amongst the pebbles were chert, tourmalinized and feldspathic grits, lydite, silicified oolite, quartz, sheared quartzite, micaceous sandstone and rhyolite. Small pellets of powder are scattered throughout the sand and may represent desilicified pebbles. At a depth of 34 ft a band of soft fine white sandstone occurs, beneath which is a coarse pebbly white sand, in turn underlain by fine even-grained white sand. In 1947 the face was 50 ft high except on the south, where ironstone was encountered at a depth of 30 ft, at which depth excavation ceased.

Under the gravel cap [43255365] south-west of Covers Farm were exposed 15 ft of yellow medium-grained sand with pebbly layers dipping 20° S., crossed by ferruginous markings parallel to the bedding in the sand below, underlain by 12 ft of yellow or buff fine even-grained sand with a small amount of clay, and containing only local irony concretions. The surface of this bed was seen to dip 7° N.

In a large pit [433540] immediately north of Covers Farm, the Folkestone Beds were originally exposed to a depth of about 125 ft, beneath some 15 ft of Gault; the upper part of the section has since been extended (Plate XA). The sand is current bedded and medium grained, white or yellow, with irregular irony patches and layers of pebbles up to ½-in diameter. Scattered throughout are pellets of white powder, and patches of black-coated sand occur at intervals. The lowest 15 ft are finer and more evenly grained than the remainder, with no pebbles, but are immediately over-lain by a well-marked pebble layer. See also p. 89 for a description of the section seen in 1953.

In 1947 this pit exposed Folkestone Beds, below the Gault, as follows: 6 in of coarse yellow and brown pebbly sand, partly iron-cemented into coarse rough sand-stone and with a top layer of hard pink flaky ironstone, on 3 ft of evenly bedded coarse pebbly buff and yellow sand, brown at the top, on 12 ft of coarse brown current-bedded sand, underlain by coarse current-bedded sand, very coarse and pebbly in places, light buff becoming paler downward, seen to 80 ft thickness.

The core and specimens of a borehole [425542] put down in 1928 at the Water Works 333 yd N.W. of Westwood Farm showed the Folkestone Beds there, 224 ft thick, to consist of 105 ft of white, yellow or buff pebbly sands, usually medium-grained with coarse and fine seams, with ironstone at top and bottom and occasionally soft sandstone, overlying 67 ft of yellow or yellowish buff, very fine sands, 31 ft of yellowish brown fine, pebbly sands and 21 ft of brown medium to fine-grained sands.

About 400 yd N.E. of Westerham railway station a disused sandpit [44955470] on the north side of Madan Road exposed, in 1947, 5 ft of evenly bedded brown sand with lenses of hard chert capping 30 ft of current-bedded yellow to brown coarse pebbly sand. The chert beds lie practically at the top of the Folkestone Beds.

A borehole at Titsey Court encountered 8 ft of white cherty rock at about the same horizon.   S.B., S.C.A.H.

About ½ mile W. of Brasted two pits [46455480] connected by a north to south cutting showed, in 1932, about 10 ft of false-bedded yellow and orange-coloured sand resting on 10 or 12 ft of evenly bedded pale yellow, clean sand in fine, medium and coarse-grained layers. In the northern pit the beds dip 11° N. and in the southern one 6° S. Excavations on the south side [47885558] of a low east to west ridge in the grounds of Combe Bank, north-east of Brasted, showed 25 to 30 ft of homogeneous white, slightly indurated sand, or soft sandstone (sandrock) which has been worked in chambers and tunnels, while on the north side of the ridge exposures [47745565] showed 5 ft of white indurated sand resting on a 1½-ft seam of hard siliceous sandstone of the Oldbury type (see p. 60), on 5 ft of yellow indurated sand with white sand below. The siliceous sandstone has been utilized as roofstone for the underground workings. The dip on the south side of the ridge is 15° S. and on the north side 15° N. In an excavation just south of the fishpond in Combe Bank grounds a section [47625550] showed 25 ft of medium grade, yellow, slightly indurated sand with harder bands near the top, and white sandrock, with gentle dip, was also to be seen in roadside exposures [48205572] in the lane east of Combe Bank House.   H.G.D.

In the lane [47335552] 500 yd E. of Brasted church some 25 ft of pale yellow medium-grade sand were seen to be capped by about 5 ft of sands with hard spicular chert bands, beneath the Gault; there is also cherty sandstone, locally cemented into laminated or more massive ironstone. Three boreholes for water less than ¼ mile N.E. of the church each proved hard beds of chert and sandstone in the top 5 to 20 ft of the Folkestone Beds. It is also notable that within a distance of some 150 yd about 3 ft of silt, with laminated clay, in the middle of the Folkestone Beds, appear to expand to some 30 ft of fine yellow and grey silt and sand, with subordinate clay, penetrated in the most north-easterly boring. Below this the sands are coarse and pebbly, then finer and ferruginous and finally rough and pebbly, with thin dark clay seams, in the basal 1 ft or more. The bottom 32 ft of Folkestone Beds in a borehole [478567] at Combebank Farm, one mile N.N.W. of Sundridge, consisted of "Dark loamy sand and thin veins of clay" (Whitaker 1908, p. 205) but in a borehole at Riverhead, 2 miles E., 121 ft of sharp ferruginous sands rested on 3 ft of glauconitic cherty sand-stone and sand at the base of the formation.

Excavations [50145643] at the border of the alluvium about ¼ mile N. of Chipstead proved thin beds of hard chert in the Folkestone Beds within 10 ft below disturbed Gault.   S.C.A.H.

At the Whitening Works [50225603], west of Chipstead Place, white sandrock beneath a small patch of river gravel has been worked in chambers about 20 ft high. In 1932, a sandpit [50435618] north of the road, opposite Chipstead Place, exposed 15 ft of homogeneous slightly compacted yellow and white sand, pale pink in the floor of the pit, and similar sandrock occurs in the road bank at the bend in the road 400 yd S.

A pit [50755530] near the base of the Folkestone Beds, 300 yd N.E. of Cold Arbor, formerly Coldharbour Farm, is 12 ft deep in yellow and brown false-bedded sand; a fault-plane with cemented sand or fault-rock, dipping 35° N.W., crosses the eastern face. False-bedded, iron-stained sand, often with coarse, pebbly layers and occasionally with irregular ironstone bands is to be seen beneath river drift, west of Riverhead, in the railway cutting south-east of the village, in pits 50 yd E. and 250 yd S. of the cutting and in a large pit [521566] once known as Marriage's, ½ mile N.E. of Riverhead, and on the south bank of the Darent, close to the alluvium. In this last pit the sand is generally coarse, with some pebbly layers; false-bedding is common, generally with a south-easterly dip of about 30°, but in the finer sand layers it is steeper than in the coarse, pebbly layers. The pebbles include both clear and opaque quartz, lydian stone and ironstone. In the south face of the pit a contemporaneous basin-shaped channel about 60 yd long and 15 ft deep is cut into the false-bedded sand, the sand filling being bedded parallel with the floor of the channel; above, the channel is cut off by drift. The mass of the sand is traversed by two sets of cross fissuring or jointing, coursing about east to west, one dipping 45° N. and the other 45° S. Trending east to west across the pit, which is upwards of 200 yd long in this direction, a hard band of fault rock 1½ ft thick, marks a reversed fault hading slightly south and stands prominently above the pit floor. In the eastern face, south of this fault, the sand is white, pale yellow or pale pink. The bedding, as shown by a pebbly layer, dips 4° to 5° S. On the north side of the fault, what appears to be the same pebbly layer dips a little more steeply south and meets the fault 8 ft below the pebbly bed on the south side. The sand on the north side is more irony, being pink to dark brown in colour, and towards the base of the pit is traversed by massive irregular ironstone seams, about 6 in thick in places, in which fossil wood has been found. These irregular layers are more or less horizontal, but against the fault they curve downwards alongside it, although they are not cut off, and no ironstone is seen on the south side; it is said that when the pit was first opened the water table stood at a higher level on the south side than on the north. Black staining in spots and horizontal layers occurs in the sand on both sides of the fault, the layers cutting across the bedding.

Exposures in the outliers on the dip slope between Brasted and Sevenoaks are infrequent. Yellow sand is seen beneath gravel in a small pit [48005454] 600 yd S.E. of Brasted Place and a small exposure [49005488] in the outlier 700 yd W. of Dryhill showed pale yellow sand, evenly bedded and false-bedded in alternating layers 1 ft or 1½ ft thick; some loamy bands in this exposure contain wisps of pipeclay. About 30 ft of false-bedded orange coloured sand below drift are exposed 600 yd S. of Cold Arbor and the entrance cutting to this pit [509548] from the north–south road at Salter's Heath is through sand with three nearly vertical bands of fault rock, trending north and south, the western one 3 ft wide and the others 1 ft 3 in and 1 ft 6 in, respectively; they are separated by 4 ft 6 in and 2 ft 6 in of sand. Around the goods yard just north of the Bat and Ball railway station at Sevenoaks, a section about 400 yd long showed near the south end a 3-ft cover of rubbly drift on 12 ft of orange-coloured sand, with cherty masses composed of sponge spicules, resting on 12 ft of reddish, rather clayey sand; the dip is about 4° N. The adjacent Greatness sand pit [535573] east of the northern end of the cutting showed, in 1930, 30 ft of markedly false-bedded sand, the foreset slopes of which dipped 30° S.E. The section showed 2 to 2½ ft of sandy soil, 6 ft of coarse false-bedded sand passing eastwards into more loamy and less false-bedded sand, 3 to 5 ft of horizontally bedded sand with thin layers of friable sandstone below, 6 ft of coarse false-bedded sand, 3 ft of horizontally bedded sand, rather loamy, and coarse false-bedded sand seen to 10 ft.

In 1947 the pit, though greatly extended, was flooded, sand being worked hydraulically. Beneath Drift (see p. 128) and re-sorted basal Gault (p. 91) 6 in to 1 ft of chert beds were seen overlying current-bedded yellow to buff coarse sand visible only up to 20 ft.

In Knole Park, south of Blackhall Farm, a small pit [54235505] showed coarse pebbly yellow and red sand with ironstone bands; in 1930, 15 ft of clean yellow and white sand were exposed in a temporary excavation [54975517] 800 yd E. of the farm. At 300 yd S.E. of the house known as Wildernesse a small pit [55085618] exposed coarse yellow false-bedded sand with thin streaks of pipeclay, and 250 yd farther east coarse pebbly brown sand with ironstone underlies a podsolized soil.

Along the face of the prominent scarp feature of Raspit Hill or Ightham Common the beds consist of ferruginous sands of more or less normal type; here and there thin white or pink clay seams an inch or so thick occur, especially towards the base of the formation, and hard cherty bands which are present near the top of the plateau-like ground of the dip-slope at Seal Chart, Raspit Hill and Oldbury Hill.

At 200 yd N.N.E. of Oak Bank, a pit [56155608] on the east side of the road showed 8 ft of false-bedded sand with impersistent chert and ironstone layers up to 3 in thick, with occasional white clay pellets, followed downwards by 6 to 8 ft of orange and white sand with tubular markings, possibly worm tubes, filled with lighter coloured sand and, in the lower part, with white clay; there are thin irregular ironstone bands and the sand contains polished grains of iron oxide. In 1924 Dr. C. J. (now Sir James) Stubblefield observed below these beds 4 ft of sand with ironstone layers, separated by a thin pink clay band from 6 ft of salmon-coloured and white sandrock.

Farther up the scarp face, chert bands bedded in a greenish clayey sand occur. About ½ mile farther east another road cutting [56655574] showed similar chert bands in loamy sand near the scarp crest, and below, up to about 20 ft of red and buff sand with ironstone bands and tubular markings, thought to be worm borings, were exposed. On the scarp crest [57415507] about 150 yd S.E. of the church, a 6-in to 9-in band of chert in laminated red and green glauconitic loamy sand with greenish clay streaks is folded into a gentle anticline and syncline, with axis trending east-by-north, while in the east side of the road a little lower down the scarp-face [57405505] were seen 8 ft of orange, red and green loamy sand with tubular markings infilled with lighter coloured sand and, at the base, a massive bed of ironstone resting on orange sand with white markings. A sandpit [57905465] ¼ mile E. of Stone Street exposed, in 1936: sand with irony chert layers 3 to 4 in thick, 8 to 10 ft; false-bedded greenish sand, 10 ft; false-bedded yellow and brown sand, coarse and pebbly with pink layers below, 45 to 50 ft; red sandrock, pebbly below, 4 ft; white false-bedded fine sand, almost sandrock, 15 ft.

From the exposures alongside a new road up the west side of Raspit Hill, ½ mile N.W. of Ivy Hatch, made about 1923, Dr. Stubblefield observed the following sequence at the top of the hill; the beds were slightly flexured: orange-coloured, loosely compacted chert rock with *Exogyra*, interbedded with grey clay seen for 7 ft; green sandstone, of Ightham Stone aspect but not silicified, and brown sand, 3 ft; sand with ironstone in irregular bands, 1 ft; yellow and green sand, 1½ ft; ferruginous chert with white kernels, 1 ft; greenish brown clayey sand, 4 ft; green and white chert passing down into brown sand with a 3-in band of ferruginous chert at base, 1 ft; brown sandrock, coarse and lighter coloured at top, sepia toned at bottom, seen to 12 ft.

G

In a pit [587545] just north-west of Ivy Hatch about 20 ft of false-bedded coarse pebbly yellow sand near the base of the Folkestone Beds were exposed beneath 3 to 8 ft of local superficial debris. Pink clay bands in the sand, and ironstone near the bottom of the exposure, were noted here by Dr. Stubblefield in 1923. The dip is about 4° N.W. and in another pit [58575492], 400 yd N.N.W., similar sand dips 3° W.

A large excavation [590556] ¾ mile S.S.W. of Ightham, also in the lower part of the formation, showed about 40 ft of evenly bedded pale yellow sand with some pebbly layers. These pebbly layers are full of silicified valves of large thick-shelled bivalves, mostly *Epicyprina harrisoni* Casey (for which this is the type-locality), *Gervillella sublanceolata* (d'Orbigny) and *Yaadia nodosa* (J. de C. Sowerby). "Most of the shells are broken and the whole deposit suggests a littoral, if not intertidal, environment" (Casey 1961, p. 542). A group of one large and several small vertical bands of fault rock trend N. 30° W. across the pit; the amount of throw of the faults is indeterminate. The dip of the strata is very gentle and is inclined toward the group of faults on either side. Near the bottom of the pit in one place a 1-ft band of pink clay occurs. Water stands in the pit bottom at the western end, presumably on impervious Sandgate Beds, which cannot be far below the floor.

The following sections at Styants Bottom are high in the Folkestone Beds sequence. A road cutting [57625650] 900 yd N. of Crown Point and on the west side of the valley showed coarse green sandstone resting on 8 to 10 ft of reddish brown loamy sand and that upon 8 ft of red and white mottled sandrock. In an old pit [578564] on the opposite side of the valley, close to the bridle path leading to Oldbury Camp, the red sandrock was seen to a thickness of 5 ft overlain by 12 ft of coarse pink sand with orange mottling and irregular ironstone bands, and with a chalcedonic cement in irregular patches. In the northern part of the pit-face the pink sand is cut by channels of coarse, false-bedded, green sand indurated in places into slabby blocks that lie parallel with the false-bedding, and near the top of the section there are spiculiferous chert bands in a clayey sand matrix. This cherty development is about 4 ft thick at the northern end of the face, the chert bands lying horizontally and occupying a basin-shaped depression or channel cut in the top of the false-bedded green sand. Only the uppermost 1 ft of the cherts oversteps the depression and extends southwards across the pink sand. The chert horizon is overlain by about 1½ ft of grey and yellow sandy clay. No Oldbury Stone occurs in this section.

In a bridle road cutting [57825649] to the north of the above exposure, the following section was exposed: rubbly chert drift, 2 ft; clayey sand with spiculiferous chert layers, 14 ft; greenish sandstone, 1½ ft; orange-coloured clayey sand with chert and wisps of greenish clayey sand, 8 ft; pale greenish yellow sandy clay with spiculiferous chert bands corresponding with the cherty horizon of the adjacent pit section, 3 ft; false-bedded orange and green sand indurated into massive blocks parallel with the false-bedding which has foreset slopes dipping 35° to 40° N. by E. (also seen in the pit section), 8 to 10 ft.

A road cutting [58205567] in the slope of the southern point of Oldbury Camp, where ground level rises above the 600 ft contour, revealed the following section, also high in the formation: buff silty sand with angular chert rubble drift, 4 to 5 ft; orange and grey sandy clay, in rough beds 3 to 4 in thick, passing irregularly into nodular spongy cherts in sand, 7 ft; coarse greenish yellow glauconitic sand, the lower 1½ ft indurated into soft green sandstone, 3 ft; orange-coloured sand and ironstone in alternating layers ½ in to 1½ in thick, 1 ft; orange-coloured and green loam, 1 ft; grey clay speckled with glauconite grains and passing in patches into soft cherts, 1 ft 4 in; green glauconitic cherty mudstone, 1 ft 3 in; beige coloured sand mottled deep orange and green, 1 ft; orange-coloured loamy sand, streaked with grey clay, 1½ ft; brown loamy sand, 1 ft 4 in; cherty ironstone, massive in places, 6 in; orange-coloured, evenly bedded sand, with white concretionary markings, 2 ft; similar sand with false-bedding, foresetting steeply south, 2½ ft; pale pink sandrock, seen to 3 ft.

North of the point where the road cutting passes through the ancient camp-ramparts the beds dip 4° N., but to the south a southerly dip of 5° and step-folds throwing the beds down southwards suggest hill-creep or cambering. Though this cutting must cross the horizon of the Ightham Stone and the Oldbury Stone, neither of these rocks was seen. Comparison with the sections to the north, described below, however, suggests a tentative correlation of the 3-ft bed of greenish yellow glauconitic sand of the above section with the Ightham Stone and of the pale pink sandrock at the base with the sandrock seen to underlie the Oldbury Stone wherever this rock is exposed. Oldbury Stone is seen near the top of the steep eastern slopes of the Camp, less than 100 yd N.E. of the cutting, and from thence is traceable for about 300 yd northwards as a bed 3 to 6 ft thick resting on white, yellow and pink sandrock. For the next 450 yd northwards along the Camp ramparts the Oldbury Stone is not seen; it may be obscured by superficial wash, but, from the bare rock outcrop that it forms both to the south and to the north it seems likely that it is absent, possibly replaced by an unconsolidated stratum. Within the Camp, to the west of this stretch in which the Oldbury Stone is not seen, is the area in which Ightham Stone was once worked for road metal. Shallow exposures in this locality, seen by Dr. Stubblefield in 1924, showed chert layers in clayey sand, resting on dark ferruginous sand which is probably at the Ightham Stone horizon; fragments of this bright green cherty rock are common constituents of the surface rubble in this area.

A sunken bridle road up the eastern slopes of the Camp showed, at the 500 ft contour [58485633], 20 ft of pink and white sandrock which is overlain by 20 ft of brown false-bedded sand. Near the top of the cutting the latter becomes very coarse and ferruginous and is cemented into large dogger-like masses which have kernels of Ightham Stone; chert beds occur above. In the uppermost part of a footpath cutting [58425639] 100 yd N. of the bridle path, similar irony sand with Ightham Stone cores was exposed, overlying 12 ft of coarse red sand, followed downward by 15 ft of sandrock, resting on loose yellow sand. In both these sections Oldbury Stone is missing but it reappears about 50 yd N. of the footpath and can be followed north-wards for about 200 yd along the eastern boundary of the Camp as a hard, compact, greyish white, quartzose sandstone up to about 6 ft thick, where, before modern depredations, it at one time was clearly seen to form the roof-stone of some of the Ightham Palaeolithic rock-shelters. The average dip is 3° N. The only other locality in which Oldbury Stone is known *in situ* is 600 yd N.E. of Styants Bottom, on the steep wooded slopes [58005681] of the valley that form the north-western boundary of the Camp. Here the quartzite again rests on sandrock, 4 ft of each being seen, but its lateral development does not extend much over 100 yd along the valley side.   H.G.D.

At Tanners Cross, north-east of Seal, the lane leading northward showed in the right bank [55695735] coarse glauconitic sand and sandstone capped by chert beds, just below the base of the Gault, dipping northwards about 10°. About 300 yd S. of Stonepitts (formerly Stonepit) glauconitic sandstone overlain by chert was again noted [56895675], here dipping 5° N., but 250 yd S.E. of Stonepitts an 8-ft section [57085684] of loamy sand with beds of chert and cherty sandstone showed horizontal bedding. Bands of hard rock recorded in borings at the Sevenoaks Waterworks pump-ing station [570576] 1 mile S.E. of Kemsing may represent these beds. Chert beds were also seen below the northern slopes of Oldbury Hill at a point [58135696] about ½ mile E.N.E. of Broomsleigh, while ¼ mile N.E. of this exposure beds of loam with sand-stone and partly cherty ironstone were noted. In all these instances the beds are stratigraphically only a little below the Gault base.

Sands of the Folkestone Beds were exposed in workings ½ mile N.E. of Ightham church, the large pit [60305745] north of the railway showing 45 ft of variable beds overlain by gravelly drift which also occurs in pockets (see p. 13). The quartzose sands are current bedded and irregularly impregnated with iron, colours varying through pink, white, buff, brown, orange and yellow. Above a well-marked ironstone layer the sand is clean and white or buff, but for a depth of 12 ft or so below it has a higher percentage of clay, though cleaner sand was found again below. The south face showed this ironstone layer about 15 ft down, with current-bedded and contorted thin ironstone in the otherwise clean sand above it. The east face showed a pink ironstone-sandstone layer, from about the same level inclined 25° N. in variable sand enclosing much highly contorted thin ironstone, mostly transverse to the main layer. The north face exposed slightly loamy beds with ironstone only near the top. Subsequent working proved the strong ironstone to be impersistent (Holmes 1937, p. 350). In a cutting between the pit and the railway there were 6 ft of white and orange sand with no ironstone.

A section low in the Folkestone Beds was exposed as follows in Dark Hill sand-pit [60455715], about ⅓ mile S.W. of Borough Green railway station: sand and sandy wash with few flints but much ironstone, 3 ft, resting on 11 ft of current-bedded brown, buff and pinkish sand with clayey partings and wisps, and with a 3-in seam of clay, 3 ft from the base, dipping north 6°; this is followed downward by 4 ft of similar sand with ironstone in seams up to 1 in thick and in contorted hollow globular masses; and this again by 4 ft of coarse white and buff sand.

A small pit [60655702] 250 yd S.E. of the above pit showed 5 ft of brown, orange and buff current-bedded sand with thin ironstone at a horizon only a few feet above the base of the Folkestone Beds.

On the north side of the railway at Borough Green railway station 15 ft of pale buff or white, current-bedded and firm sand, almost a sandrock in places, were seen in a section bordering the station yard [60955746], and similar sand was seen 150 yd to the south, in Station Road.

North of Borough Green railway station and west of Longpond the upper part of the Folkestone Beds was well exposed in four large pits. One pit [60855760] 200 yd N. of the station showed 25 ft of fine white current-bedded sand becoming brown and pink near the base and streaked with darker sand throughout. Thin tabular seams of white and buff sandstone or partly cemented sand traverse the sand at various angles and stand out on weathering. In a second pit [606578], 400 yd N.W. of the station beneath 16 ft of Gault (see p. 92) were noted 25 to 30 ft of rather coarse, current-bedded, brown, buff and pink sand, and current bedding especially in pale buff sand between persistent dull brown seams.

The deepest section was in a third pit [61005775], 350 yd N.N.E. of the station. It showed white sand (as in the first pit) underlying the darker, variable type, as follows: sandy loam, 1 ft, passing down gradually into 25 to 30 ft of brown and buff strongly current-bedded sand with large quartz grains and coarse sandy lumps, together with small pebbles of ironstone, resting with a sharp junction (dip 8° N.E.) on mainly pure white fine and firm sand, feebly current-bedded, and with oblique ½-in tabular seams of brown sand and brown and white sandstone, seen for 20 ft.

A fourth pit [610579], 100 yd N. of Longpond, showed, beneath the Gault basal beds, 2 ft of loamy sand passing down into coarse green and brown sand, overlying 35 ft of rather coarse brown and buff strongly current-bedded sand without ironstone. Descriptions of exposures at these pits have been given by Lewis Abbott (1893, p. 157), Brown (1928, p. 194) and Holmes (1937, p. 350). Drift deposits which overlie the solid are described on p. 131.

South of the railway and about 500 yd E. of Borough Green railway station a small pit [61375740] showed 10 ft of current-bedded orange sand with thick irregular masses of ironstone and ferruginous sandstone. A sandpit [61185707] near the Black Horse Inn, about 500 yd S.E. of the station, showed the following section of the lowest horizons of the Folkestone Beds: brown, buff and orange current-bedded sand with thin clayey partings and wisps of ironstone, up to 8 ft, resting on 5 ft of current-bedded sands with ironstone in 2-in bands (dipping 5° N.E.), replaced on the north by coarse orange sand which underlies fine buff sand with ironstone on the south; these beds lying on coarse pebbly sand with ironstone, slightly glauconitic, seen to 5 ft.

About 20 ft of mainly yellowish buff sand, also low in the Folkestone Beds, were seen in an old pit [62205723] 250 yd N. of Platt church. Current-bedding is not conspicuous but seams of firm sand traverse the main mass of fine loamy sand in various directions. At the top massive lenticular ironstone forms an irregular complex about 5 ft thick, on which rests a variable black layer followed upwards by irregularly bleached sand.

At Pascall's brick and tile works [620578] ¾ mile E.N.E. of Borough Green Station, sands of the Folkestone Beds were seen beneath the Gault of the northern part of the workings (see p. 93) and in pits on the south. On the north 7 ft of coarse dark brown pebbly sand, below the Gault, overlie, with a sharp junction, 12 ft of fine buff sand showing large-scale current-bedding. On the south, towards the railway, the beds are much more variable. Workings [62205765] 150 yd N.N.E. of the bridge over the railway (Brown 1928, p. 195) revealed the following developments: brown and buff current-bedded sand, exposed to a depth of 15 ft, showed incipient ironstone formation in complex patterns, then ironstone in thin seams and, farther east, in thick contorted vertical masses; the top 8 ft had a speckled and banded appearance and enclosed some vertical tabular seams of white sandstone associated with ironstone. On the north-east sand was exposed to 25 ft and ironstone was seen to be strongly developed in contorted masses (mainly vertical) at the top; fine loamy sand with contorted black layers appeared to fill hollows in the ironstone-rich sand.

In the railway cutting [62205753] just south of the above up to 30 ft of variably coloured slightly loamy firm sand showed current-bedding and strongly developed ironstone, while an adjoining sandpit [624576], 250 yd E.N.E. of the bridge over the railway, exposed 25 ft of current-bedded buff sand, with darker seams but only a little thin ironstone at the top.

East of this point the Folkestone Beds appear in a number of railway cuttings in the region of Wrotham Heath. In the cutting [62905785] north of Gallows Hill 25 ft of current-bedded, firm, slightly loamy sand enclosed irregular beds of dark sandstone and ironstone with an apparent dip of 10° E.; some thin tabular seams of sandstone traverse the sand vertically. Between this cutting and the road to Highlands other cuttings showed red sandstone masses with ironstone cores in current-bedded firm sand, the most easterly exposing 6 ft of orange and brown sand with ironstone and sandstone (in thick beds with an apparent dip 6° to 11° W.) overlying, with a sharp junction, 30 ft of variably coloured loamy sand. Topley (1875, p. 140) described the iron sandstone in this cutting as a discontinuous hard bed, in places 5 ft thick.

A pit [63055817] on the west of the main road, 250 yd S.E. of Nepicar House showed at the top up to 7 ft of pale current-bedded sand with enclosed tabular masses of ironstone and white sandstone inclined in varying directions and a silicified ironstone-sandstone complex at the base inclined up to 20° S.; beneath this were seen up to 15 ft of fairly coarse dull mottled current-bedded sand. The road cutting [632581] on the main London road at Wrotham Heath exposed 20 ft of current-bedded coloured sand passing, at the top of the hill, into 15 ft of pure white sand with 4-in brown loamy bands. The beds here show much minor variation in type.

Of three roadside pits about 200 yd E. of the inn at Wrotham Heath, the western [63555805] showed 25 ft of very strongly current-bedded pale firm sand, with darker sands and extensive contorted masses of ironstone in the south-east part. Adjoining this a smaller excavation [63585806] exposed 15 ft of buff, orange and darker sands, current-bedded and partly compacted into soft sandstone. In the eastern pit [63615805] 9 ft of brown current-bedded sand with much ironstone and dark reddish brown sandstone were seen to rest, with a sharp junction, on 20 ft of dull yellow silty loam with very fragile casts of bivalve shells. A dip of about 10° W. was noted along the plane of junction and evidently the sands seen in the exposures to the west overlie the silty bed.

In the description of this area by Brown (1941, p. 1) Pit A is that near Nepicar House and Section B the nearby road cutting. At Wrotham Heath the western pit is referred to as Pit C and the eastern as Pit D, while Pit E lies less than 100 yd N. of this last and showed 5 ft of orange sand with ironstone, but was subsequently deepened. At Pit G, about 300 yd E. of this, on the north side of the lane leading to Addington, sands of the Folkestone Beds as follows were exposed beneath a few feet of subsoil and irregularly loamy drift: orange sand, 2 ft, on brown and mottled current-bedded loam with green grains, 6 ft, this on pale current-bedded slightly loamy sand, 2 ft, with a 1-in seam of brown to black clay on the top, dipping about 11° N.

Another pit (F) excavated in more uniform sands higher in the sequence, was opened in Gate House Wood [63555825], west of the lane and about 400 yd S. of Ford Place. The section seen here, beneath 1 ft of bleached sandy soil with lumps of ironstone and a small pocket of sandy wash with flints, was a complex of ironstone and white tabular sandstone, with patches of buff and brown sand, from 7 to 8 ft thick. Sandstone, much fractured and crushed, occurs in bands up to a foot thick, and the beds are inclined south, about 35° at the base and 20° at the top, like those in the pit 250 yd S.E. of Nepicar House. Below this lie 30 ft of homogeneous light buff sand, slightly loamy.

Large sand pits (H of Brown; Plate VIB) ¼ mile N.E. of Ford Place gave a section in Gault (see p. 93) and Folkestone Beds. Beneath the Gault were seen some 30 ft of rather coarse buff sand with strong current-bedding with foresets dipping mainly to the south and east; in the south-west face of the pit the sand is in part loamy and glauconitic. A small pit [63455877] 300 yd N.W. of Ford Place exposed 3 ft of pebbly sandy wash on 4 ft of current-bedded sand near the top of the Folkestone Beds. Fine buff sand was seen at Westfield Farm [64485900] and 300 yd to the south-east [64705883], while farther south-east, in Addington Park and bordering the road to Maidstone, the sands of the lower part of the Folkestone Beds are orange, buff and grey in colour and with a fine loamy texture. About ⅓ mile E. of Westfield Farm pale loamy and rather coarse sand was proved to 20 ft [65105912].

About 700 yd E.S.E. of Trottiscliffe, along the road to Addington [64715967], an outcrop of ironstone associated with glauconitic loam appears to mark the top of the Folkestone Beds. A hard bed below the Gault recorded in boreholes within ½ mile southward consists partly of ironstone; it may be comparable with the cherty sandstone beds near Seal, etc. About 250 yd farther towards Addington a large pit [65055960] (J of Brown) showed 3 ft of rather coarse brown loamy sand, resting on 15 ft of buff and brown sands, coarse and pebbly at the base and with ironstone near top and base, overlying white and pinkish current-bedded sand, seen to 10 ft. The middle bed, variably loamy, showed an irregularly displaced base above the white sand. This sandstone in vertical seams traversed the section sparingly. The pit was subsequently extended and deepened. On the opposite side of the lane, about 150 yd S., other excavations (K) showed 30 ft of buff current-bedded sands with silver sands at the base which were also proved in trial-pits farther south.

Two small pits [660594] about ½ mile N.E. of Addington church each exposed about 12 ft of Folkestone Beds, mottled, current-bedded and loamy in patches in the pit to the north and on the east side of the lane, pale and rather coarse in the pit to the south and on the west side. Apart from these the only considerable section east of Addington is just south of Ryarsh village, where the section [670597] capped by drift of variable thickness (see p. 133), showed 17 ft of firm buff current-bedded fine sand, with darker layers including a 2-in brown seam at the base, resting on 25 ft of firm and very fine pure white sand, with brown wisps and layers near the base and a little light brown sand in one place near the top.

A borehole in the floor of the pit passed through 25 to 30 ft of very fine silica sand, followed downward by about 30 ft of coarse brown sand. The sharp junction between the buff sand and the white sand in the pit appeared to dip about 6° N.N.E. but the true bedding may lie at a lower angle.

S.C.A.H.

B. FERRUGINOUS SANDSTONE (CARSTONE) IN FOLKESTONE BEDS,
WROTHAM HEATH

(A 6827)

A. ANGULAR CHERT DRIFT ON HYTHE BEDS, LIMPSFIELD COMMON

GAULT

Between the western boundary of the map and Dunton Green the Gault was mainly seen only in minor exposures such as ponds and road banks, but it is readily traceable except where obscured by drift. The stiff grey and bluish soil of its outcrop consists for the most part of weathered Gault, with angular flints derived from drift, scattered over and embedded in it. The upper boundary is well defined on the ground surface where Upper Greensand is well developed and drift is absent; it may be indistinct where only Chloritic Marl or very attenuated Upper Greensand immediately overlie the formation. The base of the Gault is marked at outcrop by the contrast between its clay soil and the loose sandy soil of the Folkestone Beds, and by an abundance of phosphatic nodules associated with sandy clay.

In the field 850 yd S.S.E. of Titsey church, a section [41205425] for a water main exposed 4 to 5 ft of weathered bluish grey clay with small putty-coloured phosphatic nodules. Fossils collected from the 400-yd trench running east from the road indicate the *orbignyi* Subzone of the Upper Gault and include: *Inoceramus sulcatus* Parkinson, *Hysteroceras orbignyi* (Spath), *Anahoplites* cf. *planus* (Mantell), *Euhoplites spp.* indet., *Hamites intermedius* J. Sowerby (H. G. Owen, personal communication).

A large opening in Folkestone Beds (Squerryes main pit) [430540], 500 yd E.N.E. of Westwood Farm, between Westerham and Limpsfield, exposes the lowest part of the Gault, which was also formerly excavated in a pit [43105415] just to the north, now wooded, ¼ mile N.N.W. of Covers Farm. The following section was visible in 1953 (Casey 1961, p. 543):

Section of Folkestone Beds and basal Gault in Squerryes main pit, Westerham

Gault

Ft in

dentatus Zone (*benettianus* and *eodentatus* Subzones)

16. Grey, glauconitic clay with rusty streaks and iron-stained phosphatic nodules (*Lyelliceras lyelli* (d'Orbigny), *Prolyelliceras sp.*, *Beudanticeras laevigatum* (J. de C. Sowerby), *Hoplites benettianus* (J. de C. Sowerby), etc., in nodules).. .. seen 3 0

15. Grey, glauconitic clay with rafts of green sandy clay and large septarian nodules, flying to bits when tapped; rusty streaks (*Hoplites baylei* Spath, *Lyelliceras*, *Isohoplites*, etc., in nodules) 4 0

14. Blue-green sandy clay with phosphatic nodules scattered throughout and concentrated in a band at the base (*Hoplites* (*Isohoplites*) *eodentatus* Casey, *Douvilleiceras inaequinodum* (Quenstedt)) .. 3 0

mammillatum Zone (*puzosianus* Subzone)

13. Blue-green sandy clay with scattered putty-coloured phosphatic nodules 1 6

12. Band of putty-coloured nodules in matrix as above (*Otohoplites spp. nov.* in nodules) 6

mammillatum Zone (*raulinianus* Subzone)

11. As bed 13 .. .. 1 6

10. As bed 12 (*O. raulinianus* (d'Orbigny), *D. mammillatum* (Schlotheim), *B. newtoni* (Casey) in nodules, very rare) .. 4

mammillatum Zone (*floridum* Subzone)

9. Moss-green sandy clay with a line of putty-coloured nodules at base 1 10

8. Moss-green sandy clay 10

Gault

Ft in

7. Band crowded with putty-coloured nodules in matrix as above
(*D. mammillatum, B. newtoni, Cleoniceras floridum* Casey,
Protanisoceras acteon (d'Orbigny), etc., in nodules) 2–4

6. Blue-grey, dicey clay, slightly sulphurous; incipient development
of nodules at top; much glauconite and arenaceous foraminifera
in washed residue .. .. 4 ft to 5 6

5. Grey, very sandy clay with mauve and green streaks, passing up
into bed 6 .. .. 15 in to 2 6

mammillatum Zone (*kitchini* Subzone)

4. Yellow-green clayey sands 3 6

3. Band of white phosphatic nodules; densely packed; iron-stained
in places; abundant small pebbles (*Sonneratia kitchini* Spath,
C. morgani Spath, etc., in nodules, rather rare) 4–9

Folkestone Beds

?tardefurcata Zone

2. Sharp white sand, current-bedded with giant foresets; small
pebbles tending to concentrate in lines; a conspicuous 6-in
pebble-bed 20 ft above base 85 0

1. Silt Band. Buff and grey silt .. .. seen 6 0

Total about   120 0

Subsequent enlargement of the pit has enabled higher beds of the *dentatus* Zone to be seen, which comprise about 9 ft of typical bluish grey Gault clay with *Hoplites* of the group of *H. dentatus* (J. Sowerby) and *H. spathi* Breistroffer. The *kitchini* Subzone, at the very base of the Gault, occupies the north-east corner of the pit and wedges out in a westerly direction.

Casey (1961, p. 544) considers this pit as probably the most important in south-east England for studying the sequence of ammonite faunas in the *mammillatum* Zone and the lower part of the *dentatus* Zone and has illustrated many ammonites from the *floridum* Subzone of this locality in his Monograph on the Ammonoidea of the Lower Greensand (1960–6), mostly from bed 7. In addition to the species recorded by Casey in 1961, the *benettianus* Subzone of the *dentatus* Zone has here yielded *Hoplites* (*H.*) *bullatus* Spath, *H.* (*H.*) cf. *pseudodeluci* Spath, *H.* (*H.*) cf. *spathi* Breistroffer, *H.* (*H.*) *sp. nov.* aff. *cunningtoni* Spath, *Beudanticeras sanctaecrucis* Bonarelli, *Cleoniceras* (*C.*) cf. *leightonense* Spath, *Douvilleiceras sp.*, *Protanisoceras moreanum* (Buvignier), and (*fide* Mr. H. G. Owen), *Brancoceras sp.* and '*Dipoloceras*' *evansi* Spath. The underlying, poorly fossiliferous, bed 14 contains *H.* (*Isohoplites*) *eodentatus* and allied species and *Douvilleiceras inaequinodum*.

Fossiliferous Lower Gault was formerly exposed in a brickyard [44755495] closed before 1900 (London Road Brick Works) ½ mile N. of the church at Westerham (see Jukes-Browne 1900, p. 86).

In the lane cutting [47315556] 650 yd S.E. of Brasted railway station the base of the Gault is red and green sandy clay with layers of phosphatic nodules. A borehole made for the Metropolitan Water Board [47055575] 300 yd S.E. of the station, passed through 14 ft of dark bluish grey clay with phosphatic nodules followed by 5 ft of dark green sandy clay, these forming the basal beds of the Gault, which was bottomed at a depth of 64 ft. A later borehole also penetrated the lower beds of the Gault and the following summarized account of the succession observed was published by Casey (1954, p. 266): "The *dentatus* Zone was entered at about 12 feet from the surface.

The *niobe, intermedius* and *dentatus-spathi* subzones, with characteristic fossils, were all present in descending order, and appeared to be of even greater vertical thickness than at the outcrop at Dunton Green and Greatness Lane. As would be expected, the same lithological characters of the clays were seen as at those localities, though the sharp colour-change at the junction of the *niobe* and *intermedius* Subzones was not apparent. A phosphatic nodule-bed at the base of the *intermedius* Subzone, like that at Dunton Green, was encountered at a depth of 38 ft 6 in, and the underlying *dentatus-spathi* Subzone was proved to a thickness of nearly 20 ft. About 11 ft of predominantly glauconitic sandy clay, representing the basal *dentatus* Zone (*benettianus* Subzone) and the *mammillatum* Zone, then intervened before the Lower Greensand was struck at a depth of 70 ft from the surface". Further details of beds included in the *dentatus* and *mammillatum* Zones have been recorded from this borehole (Casey 1961, pp. 544–5). Coarse green sandstone (with *Entolium orbiculare*) again marks the lithological top of the Folkestone Beds.

Lobley (1880, p. 194) saw workings in Gault and brickearth at Chipstead Tile Yard [495562], now overgrown.

In the cutting for a bridge carrying the Chipstead–Chevening road over the Sevenoaks By-pass [49555680], about 6 ft of dark grey slightly micaceous and silty clay were exposed. The clay contains occasional buff-grey phosphatic nodules scattered throughout. The following fossils were collected (H. G. Owen, personal communication):

Hysteroceras aff. *varicosum* (J. de C. Sowerby), *Epihoplites sp.*, *Idiohamites spiniger* (J. Sowerby), *I.* cf. *favrinus* (Pictet). The assemblage indicates the *varicosum* Subzone.

The Dunton Green Brick and Tile Works pit [515572], ½ mile S.E. of Dunton Green (see Jukes-Browne 1900, pp. 89–90, 318, 427; Wright and Thomas 1947, pp. 315–8; Owen 1958, pp. 159–60, fig. 1), exposes some 40 ft of grey clay ranging from the middle of the *dentatus* Zone up to the *varicosum* Subzone of the *inflatum* Zone; a band of paler, calcareous and less plastic clay in the upper half shows gentle northerly dip. About 7 ft below this calcareous band and 15 ft or so below ground surface a prominent nodule-band marks the top of the Lower Gault and of the *lautus* Zone (Wright and Thomas, *loc. cit.*). Many specimens illustrated in Spath's Ammonoidea of the Gault (1923–43) came from this clay-pit. In the 15-ft deep Bat and Ball Brick and Tile Works pit [531571], about a mile E. of the Dunton Green pit, the surface clay is weathered to a depth of 4 or 5 ft to a fawn colour; below this the clay is generally bluish grey but in places it is black and has a shaly structure. Fossils are common in the clay and two bands of nodules up to 6 in thick, 6 ft apart, show a gentle northerly dip. On being washed, the clay leaves a residue of up to 3 or 4 per cent of its weight of coarse sand, mainly comminuted shells but with some fragments of limonite and well-rounded and etched grains of quartz. The face was 24 ft high when seen by Jukes-Browne (1900, p. 85) who recorded a further 20 ft of Gault as proved in a trial hole. Brown (1924, p. 80) saw in the eastern face a nodule band which he referred to the top of the Lower Gault.

At the Greatness sandpit [534574] (p. 83) the disturbed feather-edge of the Gault seems to be indicated by green clayey loams associated with the drift overlying the Folkestone Beds.

A section of over 40 ft of clay has been recorded in the workings of the Sevenoaks Brick Works Co. at Greatness Lane [536518], ¾ mile N.E. of Sevenoaks (Bat and Ball) railway station. An outline of the succession here with special reference to the foraminiferal content of the clays was given by Khan (1952), and more detailed accounts, with lists of fossils from the several beds, have been published by Milbourne (1956; 1962, pp. 439–42) and by Owen (1958, pp. 152–5, fig. 1; 1963, pp. 38–9). Milbourne (1962, pp. 437–9) also recorded a trench section in the Gault for about 800 yd northward of Child's Bridge [54525800]; in it he identified a fossiliferous phosphatic nodule bed, in a glauconitic clay matrix, with a similar bed known to occur at Wrotham and also at Aylesford and on the Maidstone By-pass, to the east of the present map. This bed is the equivalent of beds XI and XII of the Gault at Folkestone.

Near Kemsing a section at the Noah's Ark Brick Works [554581], 400 yd S. of Dyne's Farm, showed 10 ft of clay of the Upper Gault with a fauna indicative of the *varicosum* Subzone, according to Spath. Small phosphatic nodules are scattered throughout and the weathered clay is blue, but the top few feet are weathered brown and contain embedded white battered flints introduced from the surface. The following is an analysis of the clay from this pit, from a slightly weathered sample taken from about 4 ft below ground surface. Figures are percentages: SiO_2 37, Al_2O_3 16, total iron calculated as Fe_2O_3 4·3, MgO 1·3, CaO 16·0, TiO_2 0·7, P_2O_5 0·1, loss on ignition 21·4. Contains carbonate but no more than a trace of calcium sulphate. Analyst: C. O. Harvey 1937, Geological Survey Lab. No. 1032.

The base of the Gault was shown in the road bank [55695734] 120 yd N. of Tanners Cross. Here normal buff clay gives place downwards to sandy loam and sand, orange-coloured when weathered, overlying glauconitic brown loam and sand with phosphatic nodules. Beneath are sandy beds with chert marking the top of the Folkestone Beds (p. 85).

Skevington's clay pit [568578], immediately N. of Kemsing railway station, exposed some 20 ft of clay, blue when fresh, with scattered small phosphatic nodules and pale buff limy bands developed in the upper part. The rich fauna, of which the ammonites were originally identified by the late Dr. Spath and the remaining fossils by Mr. C. P. Chatwin, indicates that the beds belong to the *orbignyi* Subzone of the *inflatum* Zone. The ammonites include *Hysteroceras orbignyi* Spath, *Metaclavites trifidus* (Spath) and various species of *Euhoplites* and *Hamites*. Other common Gault fossils include the coral *Trochocyathus harveyanus* Milne Edwards and Haime, the bivalves *Inoceramus concentricus* Parkinson and *Nucula* (*Pectinucula*) *pectinata* J. Sowerby, the gastropod *Nummocalcar fittoni* (Roemer) and the belemnite *Neohibolites minimus* (Miller). Brown (1924, p. 80) also determined the presence of the subzone, and noted the occurrence near the railway line of a nodule bed separating the Upper and Lower Gault. Comparison with borehole records at Kemsing Pumping Station shows that the Lower Gault hereabouts is not more than about 40 ft thick.

Sandy yellow clay of the basement bed was struck in a number of shallow trial holes piercing the drift west of Ightham Court, where the stream here has cut down through the basal Gault, which dips gently northwards, and channelled into the underlying red sand of Folkestone Beds.

The base of the Gault and its junction with the Folkestone Beds was well seen in two sections at the sandpits north of Borough Green Station (p. 86 and Plate VIA). In the most westerly pit [607578], ¼ mile W. of Longpond, the section was as follows:

	Ft	in
Brown-weathered clay .. .. up to	4	0
Blue and purplish clay with scattered phosphatic nodules and an impersistent layer of them at the base	4	0
Highly glauconitic sandy clay, dark green and purplish in rather irregular patches; many phosphatic nodules and large sand grains	4	0
Band of phosphatic nodules set in similar sandy clay		6
Yellowish brown slightly glauconitic sandy clay passing down into	1	6
Coarse loam with clayey patches	2	0
The lowest bed passes into rather coarse sand of the Folkestone Beds.		

The section in the most northerly pit [610579], 100 yd N. of Longpond, was similar but the Gault is somewhat disturbed and with numerous pockets of coarse angular gravel (see p. 131) and the individual beds were not so well seen. Brown (1924, p. 79) states that the nodules in this area have yielded the zone fossil *Douvilleiceras mammillatum*. He includes descriptions of other sections, notably at Pascall's brick and tile works [61905765], ¾ mile E.N.E. of Borough Green railway station where Gault of the *mammillatum*, *dentatus* and *intermedius* horizons was exposed, the *mammillatum* Zone showing 2 ft of dark green highly glauconitic sandy clay with abundant phosphatic nodules and small quartz pebbles, with sand grains frequently attached to the nodules. This is underlain by 1 ft of brown glauconitic sandy clay, which passes down into coarse brown sand of the Folkestone beds (see p. 87).

Similar basal beds of the Gault occurring at the Ford Place sandpits [638589], ¾ mile S.W. of Trottiscliffe, were described by Brown (1941, p. 8) and by Casey (1961, p. 545), who deals with the palaeontology and gives the following section:

Basal Gault and Folkestone Beds exposed in sandpits at Ford Place, Wrotham, Kent

		Ft	in
Basal *dentatus* Zone			
9.	Very dark, glauconitic, sandy clay with brittle phosphatic nodules (rare *Hoplites* and *Beudanticeras laevigatum*)	3	0
mammillatum Zone (*puzosianus* Subzone)			
8.	Band of dark, gritty phosphatic nodules in a matrix of dark-green, gritty clay		6
7.	Dark-green sandy clay with scattered black-hearted, gritty phosphatic nodules	1	0
mammillatum Zone (*raulinianus* Subzone)			
6.	Band of white-skinned, dark-centred, gritty phosphatic nodules in a matrix of brown clayey sand..		4
5.	Brown-weathering, glauconitic loam with scattered, white-skinned, gritty phosphatic nodules		10
mammillatum Zone (*floridum* Subzone)			
4.	Concretionary band of whitish, friable phosphatic nodules in a matrix of reddish brown loam		2–6
3.	Grey-brown, plastic sandy clay		8
2.	Brown clayey sand with wisps of pure clay, scattered small pebbles, and incipient phosphatic nodules. Near the base a few large pebbles (up to 4 in) of micaceous siltstone (sharp junction with Folkestone Beds below)	4	0
?tardefurcata Zone			
1.	White and buff, coarse to medium grained sands, current-bedded with giant foresets .. .. seen	25	0
	Total about	35	6

The same succession, followed by a very good sequence of Lower and Upper Gault, may be seen immediately to the north [636591], where a large opening has been made to obtain clay for cement manufacture. Species of *Protohoplites*, *Otohoplites*, *Sonneratia* and *Pseudosonneratia* from the *mammillatum* Zone of these pits have been illustrated in Casey (1965, Pt. VI). Remains of the dinosaur *Camptosaurus* have also been found at this level. A nodule bed in the overlying *dentatus* Zone is full of phosphatic moulds of *Hoplites*, among which have been found species of *Oxytropidoceras*, '*Dipoloceras*', *Mojsisovicsia*, *Falloticeras*, *Brancoceras* and *Metahamites*, the whole fauna having much in common with that of the *dentatus* Zone of Escragnolles, in the Alpes Maritimes Department of France (Casey 1959, p. 207). Owen (1963, p. 38) has briefly described the *dentatus-spathi* and *intermedius* subzones of this locality and a fuller treatment of the lithological and palaeontological sequence, ranging up to the *inflatum* Zone, is given by Milbourne (1963). The Trottiscliffe waterworks borehole near this site proved a basal 5 ft of coarse, dark green, pebbly sand in a clay matrix.

Highly glauconitic sandy clay with pebbles and phosphatic nodules crops out in the road bank [64655968] about 700 yd E.S.E. of the inn at Trottiscliffe, and marks the Gault–Folkestone Beds junction.

Brown (1941, p. 8) also recorded the base of the Gault in a large pit [65055965], ⅔ mile E.S.E. of Trottiscliffe, where 3 ft of sandy clay with fossiliferous nodules, and a 1-in clay seam at the base, were seen to rest on 6 ft of brown sand with small pebbles passing down into the Folkestone Beds exposed to a depth of 40 ft (see p. 38). These occurrences mark the disturbed feather-edge of an adjacent Gault outlier. East of the pit phosphatic nodules are abundant in the soil, being derived from pockets of Gault, and from the nearby Gault outcrop.

Glauconitic clay with phosphatic nodules exposed in the road cutting [66536015] ⅓ mile W.N.W. of Ryarsh yielded a fauna including *Beudanticeras newtoni* Casey and the zone fossils of the *mammillatum* and *dentatus* Zones. Beyond this point to the eastern edge of the present area the base of the Gault is marked by the abundance of phosphatic nodules in surface soil where this is free from drift.

The upper boundary of the Gault is not well seen in the Otford and Trottiscliffe districts, being largely obscured by downwash drift.   S.C.A.H., R.C.

UPPER GREENSAND

Green clayey glauconitic sand was exposed by the side of the lane [40045459] near Limpsfield Lodge Farm, and at about the same horizon fragments of sandstone with mica and glauconite, and of malm rock with specks of mica and small grains of glauconite were noted at the pond [40115465] immediately north-east of the farm, while nearby, traces of cream-coloured malm rock with glauconite and mica from a higher horizon were seen. These fragments are hard and usually with a blue cherty core. At a higher level again green sand with mica occurs, which becomes finer, less glauconitic and more buff coloured upward as it approaches the buff marly Chalk.

Green glauconitic micaceous sandstone fragments are abundant in Broomlands Lane [41155500] about 500 yd S.W. of Pilgrims' Farm, and between the lane and the farm the base of the Upper Greensand is a sandy clay with glauconite, succeeded upwards by a cream coloured malm rock which passes into a soft, clayey, micaceous, glauconitic sandstone. Higher again there is light buff sandstone with black specks and mica, immediately underlying the Chloritic Marl (the basal bed of the Chalk), which is a buff marl with green specks. East of the farm are beds of hard green sandstone and sandstone with malmy streaks, to which may be attributed the well-defined feature on which the farm stands, while farther east [419553] the base of the Upper Greensand consists of a fine, buff, clayey sand with glauconite and the top a clayey sand with a considerable amount of glauconite, underlying the lowest bed of the Chalk, which there again is a buff marl with green specks.

About Gaysham [431554], north-west of Westerham, the top of the Gault, light grey clay weathering buff, passes upwards into a buff clay with occasional green specks, the base of the Upper Greensand, which becomes sandy upward with green specks increasing in number. Higher in the succession a green clayey sand is followed by a pale buff soft sandstone which locally grades into a yellowish brown harder sandstone and elsewhere into a micaceous glauconitic sand. The top bed is a green sand rich in glauconite.

At 1 mile N.N.E. of Westerham on the London road [44275557], the top of the Upper Greensand is a buff green sand with abundant glauconite, which passes upward into the Chloritic Marl. The base of the greensand here is a buff clayey sand.

About 150 yd W. of the watercress beds [45355589] south-east of Pilgrim House, a sandy glauconitic marl underlying the Chloritic Marl grades downwards into a clayey sand with much glauconite, this mineral dying out in a sandy clay which passes downwards into the Gault. No hard bed was found *in situ* here but the occurrence of scattered nodules of sandstone suggests one to be present. The width of outcrop here is 19 yd.   S.B.

At the spring-head [46165616] 300 yd W.S.W. of Court Lodge Farm, a small section in Chloritic Marl and Upper Greensand described by Jukes-Browne (1900, p. 91) showed underneath 3 ft of glauconitic sandy marl (Chloritic Marl) firm grey micaceous and siliceous rock, greenish at the top, passing down into a whiter malmstone. The junction with the underlying Gault was obscured but the Upper Greensand is very thin and appears to die out only a few yards to the east. From this point to Combebank Farm, north of Sundridge, the Upper Greensand is absent.

A narrow outcrop of Upper Greensand was traced from Combebank Farm through Chevening to Morants Court, just east of which the formation again practically dies out. The bed consists of highly glauconitic micaceous sand and loam, generally dark green, passing upwards into buff glauconitic Chloritic Marl and downwards into slightly glauconitic yellowish or orange-weathered Gault clay. Its maximum thickness is not more than about 12 ft along this part of the outcrop, and at Chevening it is about 5 ft, while northwards from the outcrop, a well in Chevening Park proved 3 ft of dead green sand between Chalk Marl and the sandy blue clay of the Gault.

At Shoreham, in the Darent valley a little north of the present area, a borehole at Shoreham Place passed through 10 ft of Upper Greensand (undescribed); but others, again a little north of the area, at Preston Farm, Shoreham, have indicated that if Upper Greensand is present it contains no hard beds to distinguish it clearly from Lower Chalk or Gault.

Whitaker, in manuscript notes, recorded traces of probable Upper Greensand along the outcrop up to ½ mile W. of Kemsing and it is possible that small lenses may occur there. A borehole ½ mile N. of Kemsing, at Shorehill [55545954], passed through 5 ft of loamy green sand, beneath Chloritic Marl, which may be regarded as Upper Greensand. Trenches in Childsbridge Lane, Kemsing, observed by Milbourne (1962, pp. 437–9), passed from Chloritic Marl through Upper Gault without showing any beds of Upper Greensand facies.   S.C.A.H.

Chapter V

CRETACEOUS: CHALK

GENERAL ACCOUNT

THE CHALK forms the high ground in the north-west and north of the area. Of the three main divisions recognized in the English Chalk in general, the Lower Chalk and Middle Chalk are completely represented but only the lower part of the Upper Chalk is present. At the base of the Lower Chalk occur a few feet of Chloritic Marl, a distinctive lithological type, while at the top come the *plenus* Marls. At the base of the Middle Chalk is a hard nodular bed, the Melbourn Rock, and the base of the Upper Chalk is also characterized in many places by nodular chalk. Flints are common in the Upper Chalk.

The subdivisions of the Chalk of the Sevenoaks area are as follows:

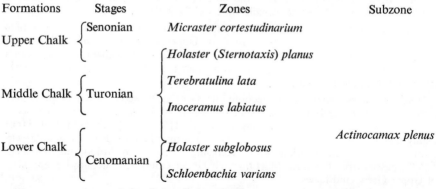

Formations	Stages	Zones	Subzone
Upper Chalk	Senonian	*Micraster cortestudinarium*	
		Holaster (Sternotaxis) planus	
Middle Chalk	Turonian	*Terebratulina lata*	
		Inoceramus labiatus	
Lower Chalk	Cenomanian		*Actinocamax plenus*
		Holaster subglobosus	
		Schloenbachia varians	

The greater part of the face of the Chalk escarpment is formed of Middle Chalk, with Upper Chalk capping it in most places, and with the Lower Chalk occupying a gentle slope at the foot which merges gradually into the Gault vale. Through a shallow wind gap at Botley Hill the Middle Chalk outcrop of the escarpment merges with that of one of the deep dry valleys of the dip slope. South of Tatsfield the Middle Chalk extends to the top of the scarp, and Upper Chalk is also absent from the edge of the escarpment east of Knockholt and north-east of Kemsing.

While neither the hard nodular chalk that occurs near the junction of the Middle and Upper Chalk nor the nodular Melbourn Rock at the base of the Middle Chalk give rise to any conspicuous topographical features, the scarp slope usually begins its conspicuous southerly flattening about the level of the Melbourn Rock, a characteristic particularly noticeable in the neighbourhood of Wrotham and Chevening. S.C.A.H.

LOWER CHALK

The thickness of the Lower Chalk increases northwards from about 190 ft as calculated from the record of the Metropolitan Water Board's well [429558] east of Tatsfield Court Farm, and about 200 ft along the escarpment generally, to 252 ft as proved in a borehole at Jewels Wood [40656075], 2½ miles N.N.W. of Tatsfield. This thickening may largely represent structural accommodation to folding along the escarpment (p. 14).

The zone of *Schloenbachia varians* includes the Chloritic Marl at the base of the Chalk, and the Chalk Marl. The Chloritic Marl (see Dines and Edmunds 1933, p. 93), nowhere clearly exposed in the district, overlies the Upper Greensand in the western part of the area where that bed is present, and elsewhere it rests on the Gault. It consists of buff or yellowish chalky marl, frequently slightly micaceous, with scattered green grains of glauconite; some authors now adopt the term 'Glauconitic Marl', but the established stratigraphical designation is everywhere used on maps and is here retained. It is never more than 6 ft thick and usually not more than 3 ft and is still thinner in the eastern part of the map area. There is a gradual upward passage into Chalk Marl.

The Chalk Marl consists essentially of greyish marly chalk and buff-grey marl, the latter strongly developed in the lower part, and exclusively in the actual basal bed. Occasional beds of marly chalk a few feet thick are comparatively hard and blocky, weathering at the surface to form lumpy bands enclosed by soft marl. West of Titsey the zone may be no more than 30 ft thick, though west of Chevening Park and at Yaldham the lithology suggests a thickness of 40 ft or more, and about ½ mile E. of Wrotham an estimate of 55 ft is made.

A paucity of exposures and of fossils does not permit a definite delimitation of the boundary between the *varians* Zone and that of the overlying zone of *Holaster subglobosus*, the chalk of which passes down very gradually into the more marly beds below. An estimated thickness of this zone including the *plenus* Marls at the top is between 135 and 160 ft.

The chalk of the *subglobosus* Zone below the *plenus* Marls has usually a massive blocky character. The upper beds are frequently dull white and the lower are yellowish grey, becoming pale and sometimes white when weathered. Occasional hard grey lumps occur but not usually in definite beds. It is in the main soft and tough but becomes brittle and flaky when much weathered. A marly tendency is sometimes evident in the lower part of the zone.

The subzone of *Actinocamax plenus* forms a constant horizon at the upper limit of the Lower Chalk but its thickness and lithology vary somewhat from place to place. For the most part it consists of yellowish grey irregularly splitting marl, with a soft lumpy texture, which passes gradually into the soft blocky chalk of the beds below; it may enclose irregular seams and lenses of firm blocky white and sometimes cream-coloured chalk. The *plenus* Marls have a maximum thickness of about 6 ft, as north of Otford, but at some other localities they are only about 2 ft or less. Recently a detailed study of their micro- and macrofauna has enabled Jefferies (1961; 1963) to correlate individual beds of the Marls over a wide area, and to study their variation in faunal content, thickness and lithology more fully than has hitherto been attempted; the subzone frequently comprises a standard succession of eight beds which exhibit shallow-water features in varying degrees.

Samples of unweathered Lower Chalk brought up in a wet condition from boreholes may appear exceptionally dark coloured and marly, as, for example, at Jewels Wood. Upon drying the material becomes paler in colour and may dry almost white.   S.C.A.H.

MIDDLE CHALK

The Middle Chalk varies in thickness from place to place between 180 and 250 ft. In the Biggin Hill district it is about 210 ft but near Titsey about 180 ft. Eastwards along the escarpment the division becomes rather thicker. Above Otford it is about 220 ft, while at Kemsing it approaches 250 ft. Near Wrotham it is from 180 ft to not more than 200 ft but may thicken again beyond this locality.

The zone of *Inoceramus labiatus* has a well-defined lower boundary provided by the Melbourn Rock. The upper limit, however, is indistinct, for there is a gradual lithological transition into the overlying zone. Moreover it is difficult to draw an exact palaeontological line between the two zones. Assuming the lithological features of the zone to be constant and distinctive a thickness of from 50 to 80 ft is indicated, which may be greatest in the Kemsing neighbourhood, where the Middle Chalk as a whole appears to be at its maximum. This includes the Melbourn Rock from 10 to 15 ft thick towards the western border of the map area, and about 25 ft at Hogtrough Hill and east of Shootfield House. The Jewels Wood borehole (p. 149) proved 11 ft of hard Melbourn Rock. In sections near Chevening and Dunton Green the upper limit of the Melbourn Rock is vaguely defined but the bed remains exceptionally thick; at Dunton Green and over the greater part of the area east of the Darent the thickness is never more than 20 ft and usually much less.

The Melbourn Rock consists of hard nodular chalk made up of small, often yellowish kernels set in irregularly fracturing chalk. The kernels vary in distinctness and development and in general diminish in proportion as the beds are traced upwards. Sometimes the nodules are little more than vaguely outlined hard cream-coloured patches; elsewhere they are small and rounded, highly concentrated and often form seams of intensely hard chalk which tend to alternate with rather more homogeneous material. The Melbourn Rock is nevertheless generally massive and uniform, lying in thick beds except for thin irregular marly partings.

The chalk of the *labiatus* Zone (and higher) above the Melbourn Rock becomes progressively less lumpy upwards but it retains a rather hard brittle and irregularly fracturing character. In the Jewels Wood borehole 89 ft of chalk above the Melbourn Rock were hard and rough or contained hard bands. The Chalk is usually white without the yellowish tendency present in the rock below.

The Zone of *Terebratulina lata* is generally from 150 to 170 ft thick but near Titsey it may be rather thinner and at Wrotham it is only about 130 ft. The greater part of the zone consists of clean white soft or rather brittle chalk, splitting squarely and with thin marly partings as well as with occasional prominent marl seams; flints, absent in the lower part, are small and scattered except in the highest beds. This lithology may persist practically to the top of the zone as, for instance, north of Botley Hill, at Chevening Park and near Kemsing, but usually the upper beds are comparable with the *planus* Zone chalk above in being less homogeneous and having seams of intensely hard nodular chalk developed at various levels. These hard bands, which are characteristically brown and yellowish and may contain numerous remains of sponges, have been noted down to about 35 ft below the top of the zone; for the most part, however, the thoroughly nodular character is confined to the top 20 ft or so. Although only scattered small flints persist to about 50 ft downward from the top of the zone, in the upper 20 to 30 ft well-defined bands of large nodular and semi-tabular flints are present and occasionally a tabular band may be developed. Marl seams have been noted at the following levels below the top of the zone: 10 to 13 ft near Dunton Green and Wrotham, 30 ft at Westerham Hill, 45 ft at Wrotham and 80 ft north-north-east of St. Clere. A definite faunal and lithological horizon at the base of the *planus* Zone of the Upper Chalk defines the top of the zone. S.C.A.H.

Upper Chalk

Not more than 120 ft of the Upper Chalk occur within the present district, and for the most part the existing thicknesses are not more than 60 or 70 ft. The zone of *Holaster planus* has an approximate thickness of 60 ft which may increase to 80 ft north-north-west of Tatsfield. Thus generally the Upper Chalk is solely that of the *planus* Zone, higher beds only occurring west of the Darent gap.

The existence of a bed of nodular chalk, with a special fauna of cephalopods and gastropods, occurring at the base of the Upper Chalk was first demonstrated by Sharpe (1855, p. 52), and was first noted in north Kent by Bensted (see Whitaker 1861, p. 170). The molluscan fauna was described by Woods (1896, 1897) who recognized that it indicated a shallowing of the Chalk sea and compared it with the fauna of comparable horizons on the continent and that of the Lower Chalk. One of the characteristic ammonites of the bed is *Hyphantoceras reussianum* (d'Orbigny). Woods suggested, therefore, that the bed, named 'Chalk Rock' by Whitaker (*in* Ramsay and others 1859, p. 296), should be placed in a *reussianum* Zone to conform with its presumed Continental equivalent.

Subsequent work (see, e.g. Davis 1926, p. 211) has shown that at least the gastropods and the bivalves once thought to be restricted to the Chalk Rock may occur at any horizon between the base of the *planus* Zone and the lowest beds of the *cortestudinarium* Zone at which hard nodular chalk is present. In the present area some of these forms may even occur in soft chalk of the *lata* Zone (Middle Chalk, see p. 106). These scattered occurrences in the Middle Chalk, in beds of quite abnormal lithology for the fauna, do not, however, invalidate the usefulness of the Chalk Rock fauna in the determination of the base-line of the Upper Chalk; for the presence of the *reussianum* fauna just below a constant marl seam provides a useful datum line, not only in distinguishing between the seams of nodular chalk in the Middle Chalk and the Upper Chalk, but also in fixing the boundary between the *lata* and *planus* zones, which is but vaguely defined by the zone-fossils themselves owing to their overlap in range. The zonal modifications in *Micraster* confirm the position of the base of the *planus* Zone as defined below. It is clear, however, that the occurrence of the *reussianum* fauna is related to a particular facies of the Chalk, and that where this facies is repeated at successive levels, as occurs in Surrey and Kent, the fauna is repeated also. Thus the term *reussianum* Zone

H

or Subzone is inopportune, since exact correlation of beds with elements of the characteristic fauna is not possible. Nevertheless, the *reussianum* fauna tends to reach its maximum development, especially in abundance and variety of cephalopods, in a restricted thickness of beds at the base of the *planus* Zone. This horizon is characterized by a lithological sequence which remains constant over a wide area and is a reliable determinant of the base of the Upper Chalk. A generalized section of these basal beds of the Upper Chalk is as follows:

	Ft
Conspicuous band of flints	—
Moderately nodular and softer chalk, sometimes with hard brown nodules	6–9
Marl seam	usually 2–4 in
Mealy and lumpy chalk with intensely hard brown and yellowish nodules variably developed, usually concentrated in the middle or lower part; moulds of gastropods, bivalves, etc; flints in the middle part frequently indefinitely scattered	5–7
Conspicuous band of flints	—

The hard rocky beds usually contain sponges and the mealy-textured chalk, in marked contrast with the nodules, is soft and easily crumbled. The hard nodules in the chalk below the marl seam may form a prominent bed of hard knobbly-weathering rock 3 ft or more thick. From this bed, or its equivalent level where the nodules are more sparingly scattered in softer chalk, a typical *reussianum* fauna has been obtained, details of which are given below. This faunal assemblage is liable to occur higher in the zone but has not been found in any of the nodular beds at lower levels, although elements of the *reussianum* fauna have been recorded in soft homogeneous Middle Chalk about 80 ft down in the *lata* Zone (see p. 106). In addition, a reduced *reussianum* assemblage including indeterminate ?gastropods, '*Cardita*' *cancellata* Woods, *Limopsis sp.* and *Subprionocyclus sp.* has been collected from soft white chalk which has been referred to the *lata* Zone at White Down near Westcott in the adjacent Reigate area.

The basal beds of the *planus* Zone show the same development in the Chatham district (see Dines and others 1954, p. 39). On the coast of Kent similar beds were described by Jukes-Browne (1903, p. 375; 1904, p. 138), who recorded two marl seams. The lower one, overlying rough lumpy chalk above a very flinty bed, corresponds to the conspicuous seam that overlies the lowest *reussianum* nodules in the Sevenoaks and Chatham areas (see Holmes 1962, p. 33).

Davis (1926, p. 211) has described the association in north-east Surrey of the *reussianum* fauna in the *planus* Zone with nodular chalk in a softer matrix immediately below a seam of marl. In the railway cutting east of Chipstead in Surrey, however, the horizon is well up in the *planus* Zone (see Dines and Edmunds 1933, p. 108), though elsewhere the bed noted with the same fauna beneath a marl seam may indeed be at a lower level. The record of a species characteristic of the *reussianum* fauna from near a lower marl seam in this cutting is therefore of interest, the section being entirely in Upper Chalk. In the Biggin Hill and Cudham district a well-defined marl seam is developed apparently some 30 to 40 ft from the base (see p. 112) and this may correspond to the higher horizon at Chipstead. Davis (1926, p. 215) drew attention to the absence of a marl seam associated with the *reussianum* fauna at Saltbox near Keston (see p. 112). This section, however, is somewhat above the basal marl seam but does not appear to extend sufficiently high in the zone to expose the horizon with the upper marl seam.

Above the characteristic basal beds, the *planus* Zone consists mostly of rough mealy and lumpy chalk with thin bands of nodules containing sponges, especially in the lower part, and some thicker and more massive hard nodular beds. Intensely hard nodular chalk is in general subordinate to softer chalk; the nodules, which may be fibrous and siliceous, are usually scattered throughout soft chalk which may form a streaky grey marly matrix. Nodular flints, many of large size, may be scattered or may be in bands (see Plate VIIIA); tabular flint also occurs. The chalk is white or somewhat yellowish. Beds in the upper part tend to be more homogeneous than the rest but somewhat flaggy.

The full thickness of the succeeding zone of *Micraster cortestudinarium* is nowhere present within the district, the maximum thickness preserved on the chalk plateau being 60 ft. Exposures are infrequent and small. The chalk is often rough and mealy, with harder patches, or it may be massive, hard and white, or smooth and yellowish. Flints are common in well-defined bands. An intensely hard bed of chalk ('Top Rock') may be present at or near the base of the zone (see p. 113 and *cf.* Davis 1926, p. 215). S.C.A.H.

DETAILS

LOWER CHALK

Chloritic Marl. Jukes-Browne (1900, p. 91) noted 3 ft of greenish glauconitic sandy marl, regarded as the base of the Chalk, overlying rocky beds of the Upper Greensand (see p. 95), at the spring head 200 yd W. of Court Lodge [465563]. In the Jewels Wood borehole were recorded 2 ft of marl with mica and glauconite which appear to be separable from Upper Greensand.

The spring head near Newbarns [51575861] is now largely overgrown, but Jukes-Browne (1903, pp. 46–7) recorded 4 to 5 ft of greyish white marly chalk containing '*Ammonites varians*' and '*Am. coupei*' underlain by buff-grey marl, and further stated that downstream from there "the glauconitic sandy marl is found with a thickness of 5 ft at least", and he was of the opinion that "the glauconitic bed passed gradually to the Gault without a well-marked base line". He also quotes Topley as recording green marly sand below the Chalk in the railway cutting south of Otford, but implies that this bed represents the base of the Chalk and not the Upper Greensand. The bed mentioned by Topley (1875, p. 153) at other localities—in a field north of Otford and near Rye House—would also appear to be the Chloritic Marl, but at Chevening Park Upper Greensand is also present.

Near the spring head [54355879] ¾ mile W. of Kemsing church, Whitaker, in a manuscript note, described a hard glauconitic bed with phosphatic nodules, regarded as the base of the Chalk Marl, underlain by soft grey marly clay.

A borehole at Shorehill [55545954], ½ mile N. of Kemsing, passed through 3 ft 6 in of 'gault with sand' referable to the Chloritic Marl, above probable Upper Greensand (p. 95).

Beds above the Chloritic Marl. An old pit [39905475] 300 yd N.W. of Limpsfield Lodge Farm showed, above talus, some 10 ft of yellowish grey and white blocky chalk with occasional denser patches and containing scattered nodules of pyrite. This chalk is almost barren of macrofossils but the lithology is suggestive of the *subglobosus* Zone. Some 10 ft of similar grey and yellowish grey chalk with pyrite nodules were exposed in another degraded pit [41355560] 400 yd N.W. of Pilgrims' Farm. It is mainly soft but occasional hard grey lumps are found, while whiter patches appear to be due to weathering. Fossils are not abundant, but *Kingena sp.*, *Exogyra obliquata* (Pulteney) [synonym: *E. conica* J. Sowerby] and *Inoceramus crippsi* Mantell occur among others. The pit is capped by some 3 ft of chalk rubble.

The sides of the lane which branches from the main road north of Titsey and directly ascends the Chalk escarpment showed a few feet of soft blocky greyish chalk of the topmost beds of the Lower Chalk. The section, though poor, continued into the Middle Chalk but no *plenus* Marls or Melbourn Rock (see p. 103) could be traced. Marly chalk is present at this horizon, however, to the west in Titsey Park and about ¾ mile E.N.E. [422558] near the Rectory.

Between Tatsfield Court Farm and Court Lodge, north of Brasted, yellowish and grey, mainly soft, marly and blocky chalk was poorly exposed in road banks and grey and massive chalk was to be seen in a degraded pit [43375562] 300 yd N.E. of Gaysham.

In an old shallow pit [46805715] 600 yd S.S.W. of Shootfield House the *plenus* Marls were exposed beneath the Melbourn Rock. About a foot of yellowish grey chalk, very marly at the top, was seen to pass down into greyish white blocky chalk.

In the lane forming the western boundary of Chevening Park and about 200 yd N. of the junction with the Pilgrims' Way, a low and weathered section [47595764] exposed Lower and Middle Chalk which included 1 to 2 ft of yellowish grey irregularly splitting marl beneath the nodular Melbourn Rock. Below this the soft greyish chalk is both blocky and marly, the latter type tending to predominate in the lower part, above the Upper Greensand. A thickness of some 50 ft of soft grey marly and blocky chalk with scattered nodules of pyrite which tends to be more marly and darker in appearance towards the base, and which belongs to the *subglobosus* Zone, was to be seen in the banks of a sunken track [48615860] in Chevening Park which, from a point about ½ mile N. of the mansion, ascends the scarp towards Knockholt Pound. Fossils are not common, only *Holaster sp.* and *Pycnodonte vesicularis* (Lamarck) being noted in the bottom 6 ft and *Inoceramus crippsi* at the top. The topmost beds pass up into a few feet of typical *plenus* Marls, yellowish grey marl with a soft lumpy texture, poorly exposed at the base of a small pit on the west side of the track. Melbourn Rock follows on above this.

West of the Darent, in the tract north of Dunton Green, blocky greyish chalk may occasionally be seen in the road banks. The spring-head section near Newbarns has been mentioned above (p. 101).

East of the Darent, Lower Chalk was formerly well shown in the railway cuttings between Shoreham and Otford. The sections were originally described by Whitaker (1872, p. 25) and are noted by Jukes-Browne (1903, p. 383) and by Dewey and others (1924, pp. 15, 17). The northern cutting is just beyond the boundary of the map area; in the southern cutting Whitaker describes a twist in the beds and a fault with northerly downthrow, south of which the beds were seen to dip 5° N., remaining horizontal on the other side. Jukes-Browne regarded the rather clayey grey chalk, over 6 ft thick, in turn passing down into lighter grey chalk, as described by Whitaker, as being the *plenus* Marls. Melbourn Rock was exposed above. Dewey and others noted a section in the southern cutting which exposed "a few feet of the *plenus* Marls resting on very hard cream coloured chalk in phacoid fragments varying in size from a quarter of an inch to a foot in length", in turn underlain by six inches of yellow marl and soft grey chalk, the beds dipping gently northward. They also recorded in the cutting [52606025] 800 yd E.N.E. of Lower Barn a fault with a downthrow of 10 ft south bringing in the Middle Chalk. Marly Lower Chalk they saw bordering the Darent hereabouts had a pinkish grey tinge.

The best section of Lower Chalk in the area was at a large pit [533594] east of Otford railway station. This showed, beneath chalky drift, up to 45 ft of massively bedded grey and yellowish grey chalk, in the main blocky and well-jointed but rather more flaggy towards the base. Some layers are rather coarse and friable and others hard, but no marl seams are present. Small pyrite nodules occur but are not common. Below this possibly 20 or 30 ft of soft grey Chalk Marl were formerly dug. Jukes-Browne (1903, pp. 49–50) described the chalk as white or nearly so at the top passing down to "grey marly chalk at the base, with *Rhynchonella mantelliana*". Dibley (1900, p. 490) states that the section exposed the *subglobosus* Zone, but the lower marly beds may well include part of the *varians* Zone.

Soft and largely marly chalk forms the gently undulating tract at the foot of the Chalk scarp eastward to Kemsing and beyond. The *plenus* Marls were exposed beneath the Melbourn Rock in a small pit [55355905] above the Pilgrims' Way about 400 yd N.W. of Kemsing church. At the top is a yellow marl seam 2 in thick with the chalk immediately below in places fissile and marly. Beneath this, up to 10 ft of soft blocky grey chalk with thin irregular marly partings were exposed where small faults displace upwards the overlying Melbourn Rock. The upper part of this chalk, which is classed with the *plenus* Marls, yielded ossicles of *Crateraster quinqueloba* (Goldfuss), *Anomia papyracea* d'Orbigny and *Metoicoceras pontieri* Leriche. Osborne White (*in* Bennett 1907, p. 120) records grey or greenish grey laminated marl here.

Between Kemsing and Wrotham greyish chalk was seen at intervals in road cuttings. Just north of Yaldham grey chalk passes down into more marly chalk, and similarly about ¼ mile W. of Wrotham on the road to Yaldham, beneath a few feet of chalky drift, soft grey fissile but rather lumpy chalk passes down into buff marl [60685908]. A thin capping of Chalk Marl on the Gault forms a small outlier [565583] south-west of Heaverham.

In a shallow excavation [61255937] 250 yd N.N.E. of Wrotham church 3 ft of hard lumpy chalk with intensely hard yellowish kernels (Melbourn Rock) overlie Lower Chalk with an indistinct 2-in marl seam passing down into blocky grey chalk. A little less than ½ mile E. of Wrotham a small cutting [619591] for the by-pass (Holmes 1937, p. 352) showed about 10 ft of rather hard partly fissile and marly grey chalk with some pyrite and many fossils. *Kingena sp.*, *Orbirhynchia mantelliana* (J. de C. Sowerby) and *Grasirhynchia martini* (Mantell) are particularly common; other fossils include *Terebratulina nodulosa* Etheridge, *Aequipecten beaveri* (J. Sowerby), and *Sciponoceras baculoides* (Mantell), together with serpulids and fragmentary *Acanthoceras*. None of these is individually diagnostic, but both this assemblage and the lithology are suggestive of the *varians* Zone.

Osborne White (*in* Bennett 1907, p. 120) mentions that the greyish buff argillaceous beds of the Chalk Marl are exposed occasionally in the sides of the lanes, and about the spring heads, on the lower slopes of the escarpment to the west and north-east of Wrotham. Eastward of Wrotham the Chalk Marl forms small spurs extending from the Chalk escarpment. Its presence is indicated in the spring heads though it is almost entirely obscured by chalky drift. An extensive spread of chalky and flinty drift also masks the beds near Trottiscliffe where marly chalk is rarely to be seen at the surface (see p. 131).   S.C.A.H.

MIDDLE CHALK

The best exposures of Middle Chalk are situated on the escarpment face. Hard lumpy chalk of the Melbourn Rock horizon may be traced across Titsey Park and was poorly exposed in the cutting [40775535] of the main road from Titsey to Botley Hill. In the lane which branches from this road on the east and ascends the escarpment directly north of Titsey church, however, soft blocky grey and white chalk, practically indistinguishable from Lower Chalk, forms the local base of the Middle Chalk [40955533]. Higher up this lane the Middle Chalk was seen to contain indistinct nodular beds towards its top. Hard nodular Melbourn Rock may be traced about ¼ mile both west and east of Tatsfield Court Farm [425558]. Above it occasional small exposures in the Downs for about a mile E. of Tatsfield church show clean white soft and brittle chalk of the *lata* Zone. For the most part this tends to be flaggy, but in places it splits into small cuboidal blocks. An old pit [41905605] 200 yd E.S.E. of Tatsfield church, showed 10 ft of firm massive chalk with some lumps containing sponges. The section evidently exposed part of the *lata* Zone but the only determinable fossil yielded was *Plicatula barroisi* Peron.

About 50 ft of the upper beds of the *lata* Zone were seen in a large pit [44005615] 150 yd N. of The Mount. The section did not quite reach to the Upper Chalk. At the top there are a few flints, and rather harder occasionally nodular courses run through the mainly brittle chalk. Several seams of fissile marly chalk occur, one in particular about 15 ft from the top. In the lower part of the pit the chalk is soft, white and without flints. Fossils are not common but *Inoceramus lamarcki* Parkinson and *Romaniceras sp.* were obtained. Dibley (1900, p. 489) refers to this pit and lists a number of fossils including *Conulus subrotundus* Mantell), but there is no doubt that the chalk does not extend below the *lata* Zone.

On the opposite side of the road, about 400 yd N.E. of the pit [442564], the junction of the Upper and Middle Chalk was traced in another old pit (see p. 108). In 1939 only about 5 ft of mealy and lumpy chalk of the top of the *lata* Zone were visible, but Jukes-Browne (1903, p. 383) described 35 ft of soft white chalk with three well-marked layers of flint, assigning all to the *lata* Zone. It is probable, however, that part of this is in Upper Chalk as now understood. Besides *Terebratulina sp.* the following fossils have been obtained from the 5 ft definitely of the *lata* Zone: *Holaster (Sternotaxis) planus* (Mantell), *Limatula wintonensis* (Woods) and *Plicatula barroisi*.

The Melbourn Rock is indicated just west of Pilgrim House by hard lumpy chalk with yellowish kernels. An overgrown pit [44955653] 200 yd N. showed a band of intensely hard spongiferous nodules in rather hard white flinty chalk of the *lata* Zone containing *Spondylus latus* (J. Sowerby) and *Inoceramus labiatus* (Schlotheim) *latus* (J. de C. Sowerby).

About 400 yd W.N.W. of Court Lodge, a small pit [46155640], now considerably enlarged, noted by Jukes-Browne (1903, p. 383), showed 13 ft of massive beds of hard nodular Melbourn Rock with abundant *Orbirhynchia cuvieri* (d'Orbigny) and *Inoceramus labiatus*. These are also common in an overgrown exposure [46035684] about 450 yd N.N.W. of the preceding, and nodular rock with fragmentary *Inoceramus* may be seen in an old shallow pit [46805713] 600 yd S.S.W. of Shootfield House (see also p. 102).

South and south-east of Brasted Hill Farm [46255740] soft and brittle white chalk of the *lata* Zone containing *Inoceramus lamarcki* may be traced downwards into the more lumpy and irregularly fracturing chalk of the *labiatus* Zone.

Of two adjacent old pits [475577] about 500 yd E. of Shootfield House the upper one shows some 15 ft of white chalk of variable texture, apparently of the upper part of the *labiatus* Zone, yielding towards the top *Conulus subrotundus* and *Terebratulina striatula* (J. de C. Sowerby *non* Mantell). A layer of intensely hard silicified chalk a few inches thick was noted here. The lower pit mentioned by Jukes-Browne (1903, p. 383) exposed 20 ft of hard massive nodular Melbourn Rock grading into 5 ft of weathered lumpy chalk at the top. *Orbirhynchia cuvieri* and *Inoceramus labiatus* are common. The Middle Chalk exposed in the lane ascending the escarpment and forming the western boundary of Chevening Park consists of about 100 ft of chalk, hard and nodular in the lower part but becoming predominantly soft and brittle above. *Conulus subrotundus, Concinnithyris protobesa* Sahni and *Orbirhynchia cuvieri* were obtained from near the top of this section.

Along the northern edge of Chevening Park small overgrown pits [47585875, 47905894 and 48025899] showed soft white chalk near the top of the *lata* Zone. Among other fossils, *Cidaris serrifera* Forbes, *Terebratulina lata* Etheridge and *Inoceramus lamarcki* were collected. The *labiatus* Zone is present in two old pits [48555890] on opposite sides of the path leading from Chevening to Knockholt Pound, about ½ mile W. of Old Star House, formerly The Beacon. That on the western side showed 20 ft of hard nodular Melbourn Rock, with *O. cuvieri*, in massive courses. The nodules are of various sizes but are mainly in the form of small round kernels, often associated with cream-coloured patches in the chalk. Thin irregular marly partings were seen towards the top. The larger pit on the opposite side of the footpath is much degraded but it exposed 10 ft of hard nodular chalk and softer chalk some distance above.

Middle Chalk low in the *lata* Zone was formerly quarried just north-east of Old Star House [495586]. The chalk is white and contains very few flints, and from the uppermost 6 ft *Terebratulina sp.*, *Inoceramus inconstans* Woods and *I. lamarcki* have been obtained.

The Dunton Green Lime Works pit [502591] about 700 yd N.W. of Mount Pleasant offers one of the best sections of Middle and Upper Chalk in the district. Beneath nodular chalk of the *planus* Zone (see p. 109) the section of the Middle Chalk was as follows: white chalk, with some flints, and of variable hardness, in part weathering into nodular chalk with soft mealy patches, 10 ft; marl seam, 3 in; mainly soft marly chalk, slightly lumpy at the extreme top, partly obscured, 20 ft; blocky, irregularly jointed and slickensided, greyish white chalk, practically flintless and with impersistent seams of firm marly chalk, 30 ft.

The nodular character of the uppermost bed was seen to advantage in a small subsidiary weathered face east of the main quarry. Notable is the abundance of *Holaster* (*Sternotaxis*) *planus* together with *Terebratulina lata*. Sponges are numerous and also ossicles of *Isocrinus* and *Spondylus spinosus* (J. Sowerby). From the top three feet *Inoceramus sp.* and *Plicatula barroisi* were obtained; '*Ventriculites*' and *Spondylus* are particularly common at this level. In the lowest 30 ft fossils are not abundant, but teeth of *Anacorax falcatus* (Agassiz) and of *Ptychodus* were obtained.

Dibley (1900, p. 490) suggested that possibly the upper part of the '*cuvieri*', i.e. *labiatus* Zone, might be present in this pit, but detailed collecting shows that the *lata* Zone is overlain by some 30 ft of the *planus* Zone chalk and that its lower limit is probably some way below the present floor.

Along the main London road leading from Sevenoaks via Polhill there were occasional small exposures of Middle Chalk in the cuttings. East-south-east of the Dunton Green Lime Works the chalk is rough and lumpy and splits irregularly, a type of lithology generally associated with the *labiatus* Zone, but at higher horizons it becomes smoother and more homogeneous and contains flints. *Inoceramus labiatus* is fairly common in these cuttings. An old pit [50475986] 500 yd S. of the Polhill Arms Inn showed 15 ft of soft white chalk from which *Inoceramus inconstans*, *I. labiatus* and *I. lamarcki* were obtained near the top; in the cutting of the lane that ascends the hill immediately north of this, 40 ft of similar chalk above it yielded *I. inconstans*. These exposures are in the *lata* Zone.

About 750 yd N. of Sepham Farm rather soft clean white chalk, high up in the Middle Chalk, was seen in a large degraded pit [510606], on the northern boundary of the map. Dewey and others (1924, p. 17) described this and noted also that on the eastern side of the valley of the Darent, where the Melbourn Rock gives rise to a marked change of slope, the hard nodules of the rock lie scattered over the fields. The rock was formerly seen in nearby road and railway cuttings, and the beds were described in the railway cuttings between Shoreham and Otford both by Whitaker (1872, p. 25) and by Jukes-Browne (1903, p. 383), who noted that a bed 5 ft thick constituted the base of the Melbourn Rock.

Eastward from Otford the Melbourn Rock may be traced a little above the Pilgrims' Way to beyond Kemsing, but the only good section was in the small pit [55355905] (see also p. 103) about 400 yd N.W. of Kemsing church. Here the hard nodular beds noted by Osborne White (*in* Bennett 1907, p. 120) overlying the Lower Chalk consist of massive courses of nodular rock tending to alternate with bands of hard but more homogeneous massive chalk, and to have at the base a 6-in bed of particularly well-developed nodular rock consisting of many small round closely packed kernels. The displacement of this basal nodular bed and the underlying *plenus* Marls by three small faults is a conspicuous feature; up to 6 ft of the Melbourn Rock were exposed, but its total thickness is undoubtedly more than this.

Higher beds of the Middle Chalk are poorly exposed on the escarpment in lane cuttings between Otford and Kemsing and consist for the most part of soft and brittle chalk without flints. Occasional lumpy beds occur towards the top of the division. The chalk becomes harder and fractures more irregularly as it passes downwards into the Melbourn Rock. *Holaster (Sternotaxis) planus* and *Inoceramus labiatus latus* were obtained from near the top of the *lata* Zone 100 yd N.N.W. of Otford Court, formerly Beechy Lees [543596].

About 15 ft of soft white chalk of the *lata* Zone are present at the top of a disused pit [549594] a little over ½ mile N.W. of Kemsing church. The chalk is blocky, with occasional marly partings and a few thick-rinded flints. Fossils are rather scarce but *Holaster (Sternotaxis) planus, Terebratulina lata* and *Spondylus spinosus* were obtained about 10 ft from the top of the quarry face. Osborne White also lists a number of fossils found towards the top of this pit by F. J. Bennett, including *Micraster corbovis?* Forbes. Leach (1921, p. 38) states that the pit ('Shore Hill quarry') was thought by G. W. Young to lie in the *planus* Zone.

Eastward of this locality soft and brittle clean white chalk forming the upper part of the Chalk escarpment (above 550 ft O.D.), though poorly exposed, yields abundant *Terebratulina lata* and *Inoceramus lamarcki*. The *lata* Zone, in fact, has no Upper Chalk capping it for some distance along the scarp near Cotman's Ash [566597]. The lower part of the Middle Chalk was only exposed in one locality between St. Clere and Wrotham, in an overgrown quarry [58705925] 550 yd N. of Yaldham, where Melbourn Rock, consisting of a foot or two of firm chalk with small hard kernels, was noted.

Some 550 yd N.N.E. of St. Clere a pit [57855960] showed about 25 ft of soft blocky white chalk in massive courses, with thin irregular marly partings but practically no flints or hard patches. Fossils found here include *Parasmilia sp., Holaster (Sternotaxis) planus* and *Peroniaster nasutulus* (Sorignet). In addition *Lewesiceras* cf. *mantelli* Wright and Wright and a poorly preserved and crushed heteromorph (*Hyphantoceras?*) were collected 9 ft above and 7 ft below the marl seam respectively. The horizon is undoubtedly in the *lata* Zone (about 80 ft down) though these forms are generally associated with the *reussianum* fauna at the base of the *planus* Zone. Eastwards towards Wrotham similar chalk, but partly flaggy and tending to include lumpy horizons towards the top, was exposed sporadically in old pits and lane cuttings at levels between 550 and 650 ft O.D. Varieties of *Inoceramus lamarcki* are common.

In the neighbourhood of Wrotham typical Melbourn Rock may be traced from an exposure 250 yd N.N.E. of Wrotham church (see p. 103) in an E.N.E. direction past Hognore Farm and below the Pilgrim's Road to the edge of the map area.

North-west of Wrotham the full thickness of the *lata* Zone, about 130 ft, was exposed in the cuttings bordering the main London road. The top 10 ft or so [60505984] consist of mealy and lumpy chalk with scattered hard nodules and nodular flints, similar to the overlying basal beds of the *planus* Zone (see p. 110). Moreover, at the base of this chalk is a nodular band of hard yellowish spongiferous rock about 1 ft 6 in thick, yielding *Spondylus spinosus* and *Inoceramus lamarcki* but not, apparently, containing any of the gastropods, etc. characterizing the similar bed at the base of the *planus* Zone (see p. 99). Below this again come about 35 ft of mainly soft white chalk which weathers into large rounded masses; flints are present only near the top. A marl seam 3 to 4 in thick was noted a few feet below the rock-band and some obscure marly partings apparently crossing the bedding planes. Occasional small faults are present. At a point ½ mile N.W. of Wrotham church continuity of the section was interrupted, but below this, adjoining the pit described below, were seen 30 ft of soft blocky chalk with an impersistent band of fissile chalk in the middle. Below the pit, about 500 yd N.N.W. of the church, 10 ft of brittle and rather hard white chalk with

Terebratulina lata and *Conulus* lie at about the junction of the *lata* and *labiatus* zones [610596].

The chalk pit [60805975] 700 yd N.N.W. of Wrotham church exposed about 70 ft of the *lata* Zone. This consists of soft blocky white chalk in massive courses, with numerous thin marly partings and one conspicuous marl seam, 2 to 4 in thick, about 40 ft above the quarry floor and displaced here and there by small faults. Small grey-rinded flints occur practically only above the marl seam. The chalk of the lower part of the pit is on the whole more blocky than that of the upper part, and other impersistent fissile bands occur. At the top of the pit are the nodular rock band 10 ft from the top of the zone and the marl seam 3 ft 6 in below it, both noted in the London road cuttings. The zone-fossil is fairly common in the lower part of the pit and also near the top. Other fossils include *Orbirhynchia sp.* from above the conspicuous marl seam, *Micraster corbovis* from 30 ft below the marl seam, *Inoceramus inconstans* and *I. lamarcki* from the lowest 30 ft. This section is referred to by Osborne White (*in* Bennett 1907, p. 121), with a list of fossils, and again by Holmes (1937, p. 352).

About ½ mile N.E. of Wrotham church two old pits adjacent to the Gravesend road [61705975] exposed chalk of the *lata* Zone (Jukes-Browne 1903, p. 382). Of these the lower one east of and below the road, is now much overgrown, but 30 ft of soft white chalk splitting rather irregularly, and without flints, yielded *Inoceramus lamarcki* and *Plicatula barroisi*. In the higher pit, below the basal beds of the *planus* Zone (see p. 110), are about 60 ft of *lata* Zone chalk, of which the top 20 ft are rather lumpy with two thin bands of intensely hard nodules with sponges in the middle, and about 7 ft apart. There are also scattered hard brown lumps in the lowest 5 ft of this chalk. In addition to several bands of nodular flints, a conspicuous marl seam 3 to 4 in thick lies 2 ft below the upper band of nodular chalk and is displaced 5 ft by a small fault. In the soft blocky chalk immediately below, flints become scarce except for a tabular band towards the top. *Micraster corbovis* is common, associated with *Holaster* (*Sternotaxis*) *planus*, in the 5 ft of chalk below the marl seam and *Terebratulina lata* is abundant 12 ft below it. Osborne White (*in* Bennett 1907, p. 120) states that the upper limit of the Middle Chalk may be inferred as a result of collecting from a group of 'passage-beds' several feet thick, but a distinct dividing line between the Upper and Middle Chalk is now recognized.

In the dry valley north of Botley Hill, Middle Chalk was exposed in a small pit [40085588] 500 yd N.N.E. of the crossroads on the hill, where smooth soft white chalk, near the top of the *lata* Zone, with slightly marly seams but no flints, has yielded *Peroniaster nasutulus* and *Goniomya sp. nov.* Northwards soft white powdery chalk typical of the *lata* Zone may be traced where the valley sides are bare. Chalk of this zone was also seen in the valley to the west of Biggin Hill but, though clean and white, it tends to be brittle and fractures irregularly; at the extreme top occur hard nodules similar to those in the overlying *planus* Zone. In the Cudham Valley, *lata* Zone chalk occurs in a small pit [44265983] 280 yd W.S.W. of Cudham church. This pit showed 5 ft of brittle and soft white chalk without flints. Near Knockholt, owing to minor rolling of the Chalk strata and flattening of the dip mentioned on p. 14, the valleys have been cut in Middle Chalk of the *lata* Zone as far northwards as Blueberry Farm [474597], where Upper Chalk is brought in by the dip of the north limb of the local flexure. In an old pit [46555900] 250 yd W.N.W. of Knockholt church 5 ft of firm blocky white chalk with a clean fracture, hard and without flints were seen beneath Clay-with-flints. Fossils include '*Cidaris*' *serrifera*, *Terebratulina lata* and *Inoceramus cuvieri* (J. de C. Sowerby). Soft white chalk was seen 200 yd N. of Shelleys [46355940].

East of the Darent, 8 ft of massive soft greyish white chalk without flints of the *lata* Zone, yielding *Micraster corbovis* and sponges, were noted in an old pit [54396039] 250 yd N.E. of Paine's Farm. This chalk had been previously placed in the *planus* Zone by Dewey and others (1924, p. 40). S.C.A.H.

UPPER CHALK

Holaster (Sternotaxis) planus Zone. Chalk of this zone caps the escarpment in most places. East of Botley Hill intensely hard yellow-hearted and partly siliceous nodules were noted and farther east, where the Upper Chalk re-emerges from beneath the Clay-with-flints near Betsom's Hill, very hard nodules with sponges occur.

In the old pit [442564] 600 yd S.S.W. of Gray's Farm (see p. 104) the section was as follows:

		Ft	in
	Platy-weathering chalk	4	0
	Brittle to rather lumpy chalk, with flints scattered and in indistinct bands	4	0
	Conspicuous band of semi-tabular flint		
planus	Chalk as above flint band	6	0
Zone	Marl seam	0	2
	Mealy and lumpy chalk with, towards the base, a 6-in wide indistinct band of irregular nodular flints; associated with these in particular are brown and yellowish hard lumps with sponges and moulds of gastropods, etc.	2	6
	Mealy and lumpy chalk	3	0
	Band of large nodular to semi-tabular flints		
lata Zone	Mealy and lumpy chalk, mainly soft, with three lines of nodular flints	5	0

Micraster praecursor Rowe[1] is fairly common about 11 ft above the marl seam, and *M. corbovis* also occurs. In the lumpy chalk below the marl seam the *reussianum* fauna with *Cuspidaria caudata* (Nilsson), '*Turbo*' *gemmatus* J. de C. Sowerby, and other fossils occur; evolutional features in the examples of *Micraster* found corroborate the placing of the zonal boundary just below this level. *Holaster (Sternotaxis) planus* is found in the *planus* and in the *lata* chalk; other fossils from the *planus* beds include *Gauthieria radiata* (Sorignet), *Terebratulina lata*, *Plicatula barroisi* and *Lewesiceras mantelli*.

Higher beds in the zone were exposed in an old pit [44485666] 300 yd S. of Gray's Farm, where 18 ft of weathered soft chalk with many scattered nodular flints have yielded many fossils, including *Sporadoscinia* cf. *alcyonoides* (Mantell), *Holaster (Sternotaxis) planus*, *Micraster praecursor*, and, towards the base, *Cretirhynchia minor* Pettitt, *Gibbithyris spp.* and *Scaphites geinitzi* d'Orbigny. Chalk not far above the base of the zone was seen in a small pit [44925666] 450 yd S.W. of Silversted; this showed about 7 ft of soft and brittle flaggy chalk with nodular flints and containing the zone-fossil, *Micraster praecursor* typical of the zone and numerous brachiopods including *Cretirhynchia cuneiformis* Pettitt, *C. minor*, *Gibbithyris semiglobosa* (J. Sowerby) and *Orbirhynchia dispansa* Pettitt. Between this locality and Knockholt, where the gap

---

[1]The conservative classification of *Micraster* used in this memoir does not imply rejection of the results of recent biometric studies, which suggest that *M. praecursor* Rowe and *M. leskei* (Desmoulins) are variants of *M. cortestudinarium* (Goldfuss) (see Kermack 1954 and Nichols 1959).

(A 6830)

A. Chalk overlain by Clay-with-flints; north of Westerham

PLATE VIII

B. Head resting on Folkestone Beds, Borough Green

(A 7126)

occurs in the Upper Chalk capping of the escarpment, mealy and hard nodular chalk referable to the lower part of the *planus* Zone was occasionally exposed. In particular, 600 yd W. of Sundridge Hill *Ventriculites chonoides* (Mantell), *Inoceramus costellatus* Woods and *Scaphites geinitzi* were obtained from very hard yellow-hearted nodules. Similar chalk in an overgrown pit [49155914] 700 yd N.N.W. of Old Star House yielded many fossils, including *Inoceramus costellatus*, '*Turbo*' *gemmatus*, *Micraster praecursor* and sponges.

The Dunton Green Lime Works Quarry [502591], 700 yd N.W. of Mount Pleasant, exposed some 35 ft of chalk of the *planus* Zone above the Middle Chalk (see p. 105). The section of the beds seen in the upper part of the quarry is approximately as follows: massively bedded chalk, with flints in ill-defined bands, 15 ft; white chalk with flints in three well-marked bands, one at the top followed by two below at intervals of 2 ft and 9 ft; band of hard nodules with sponges, 1 to 2 ft; moderately lumpy chalk, 1 ft; marl seam, 4 to 6 in; moderately lumpy chalk, 1 ft; indistinct band of nodular flints on moderately lumpy chalk, rather brittle, 3 ft; on nodular chalk, partly soft and mealy; intensely hard nodules with sponges and bivalve moulds etc., 3 ft; band of large nodular flints, with Middle Chalk beds below (see p. 105).

The nodular bed at the base of the zone has yielded *Guettardiscyphia sp.*, *Dentalium turoniense* Woods, *Inoceramus costellatus*, *Austiniceras? curvatisulcatum* Chatwin and Withers and *Scaphites geinitzi*, amongst other forms. The chalk above this contains an abundant fauna of which the most important member is *Micraster praecursor*, the zonal modifications of which indicate the horizon of the beds. The zone-fossil is common throughout. Reference to these beds has been made by Seeley (1891, p. 107), Davies (1914, p. 119) and Herries (1920, p. 220).

The nodular horizon at the base of the zone was also noted 300 yd W. of the pit; this yielded several specimens of *Inoceramus costellatus* and *Subprionocyclus neptuni* (Geinitz), both characteristic of the *reussianum* fauna, in addition to *Subprionocyclus* cf. *hitchinensis* (Billinghurst), *Sciponoceras sp.* and *Scaphites* (*Otoscaphites*) cf. *auritus* Schlüter. From chalk higher in the Zone at a shallow pit in the wood above the Lime Works *Echinocorys scutata* Leske, *Micraster praecursor* and the zone-fossil were obtained.

In the lane cutting [505602] about 100 yd S.S.E. of the Polhill Arms Inn hard nodular chalk at the base of the *planus* Zone has yielded sponges including *Tremabolites perforatus* (T. Smith) and also *Calliostoma schlueteri* (Woods). This rock chalk was recorded by Dewey and others (1924, pp. 27, 39).

On the east side of the Darent valley nodular chalk at the base of the Upper Chalk is to be traced along the escarpment above Otford and Kemsing. A small chalk pit [55035955] on the edge of the Clay-with-flints capping of the plateau and situated about ½ mile E. of Otford Court formerly showed the chalk to contain many solution pipes (see p. 127). A few feet of soft chalk with hard lumps and nodular flints remained unobscured in 1937 and from it were obtained gastropod moulds, *Scaphites* and *Dentalium turoniense* together with *Orbirhynchia dispansa*, *Terebratulina lata* and *Inoceramus lamarcki apicalis* Woods, which is quite commonly found in the lower part of the *planus* Zone in the district.

Beyond the gap in the Upper Chalk outcrop in the Cotman's Ash neighbourhood [567600], chalk of the *planus* Zone occurs at the edge of the scarp above St. Clere. Hard yellowish and brown-stained nodules with *Craticularia fittoni* (Mantell) and an indeterminable scaphitid were noted some 400 yd N.N.W. [58075977] of the northern right-angle bend in the Wrotham road, and *planus* chalk in an overgrown pit [593598] 700 yd E.S.E. of Old Terry's Lodge. In a sunken track up the scarp to the east of this the following beds were exposed: soft chalk with a gritty feel, and with flints and *Holaster* (*Sternotaxis*) *planus* and *Inoceramus* cf. *inconstans*, etc. at about 725 ft O.D. [59676002], similar chalk with hard patches yielding *Holaster planus* and *Cuspidaria sp.* at about 700 ft O.D. and intensely hard nodular chalk set in softer chalk at about 690 ft O.D. [59066000]. This latter bed contains gastropods and bi-valves of the *reussianum* fauna, here at the base of the zone, including '*Turbo*' *gemmatus* var *a* Woods (this is an undescribed species unrelated to '*T*.' *gemmatus*), *Inoceramus costellatus* and indeterminable scaphitids.

The road cutting north-west of Wrotham exposed, near the top of the hill, about 20 ft of *planus* chalk and the junction with the underlying *lata* Zone chalk (see p. 106 and Holmes 1937, p. 352). About 200 yd below the bend in the road at the summit it is obscured by hill wash. East of this the basal 5 to 6 ft of the *planus* Zone cap the section for about 100 yd, dipping E. very slightly towards a small fault. West of the obscured portion the complete section [602599] was as follows:

		Ft
planus Zone	Lumpy dull white and yellowish chalk with bands of large flints, one at the base having a thin band of hard yellow nodules with sponges immediately below it	10
	Moderately soft white chalk with gritty feel and with hard patches but very few flints	9
	Thin marl seam	
	Chalk as above	2–3
	Intensely hard nodular rock, rather irregular and impersistent, with many yellow-hearted nodules containing moulds of gastropods and sponges	3–4
	Band of large nodular flints	
lata Zone	Soft chalk with occasional hard nodules	1–2

Holaster (Sternotaxis) planus is common in the chalk above the basal nodular bed of the zone and is associated with *Micraster corbovis* and *M. leskei* (Desmoulins) with the sutured interporiferous areas characteristic of the *planus* Zone. *Terebratulina lata* was obtained in the top 10 ft of the section.

The basal rock band contains *Ventriculites chonoides*, *Eurete sp.*, *Parasmilia sp.* and *Proliserpula (Proliserpula) ampullacea* (J. de C. Sowerby); fossils typical of the *reussianum* fauna including *Guettardiscyphia stellata* (Michelin), '*Turbo*' *gemmatus*, *Scaphites geinitzi* and *Inoceramus costellatus*.

On the scarp slope [61305975], 650 yd N.N.E. of Wrotham church, lumpy chalk with flints low in the *planus* Zone yielded *Holaster (Sternotaxis) planus*, *Micraster prae-cursor*, *Orbirhynchia reedensis* (Etheridge), '*Turbo*' *gemmatus* and *Scaphites sp.* from a band 4 to 6 ft down and *Inoceramus costellatus* together with common *Spondylus spinosus*, about 12 ft lower down in the section.

The Middle Chalk of the large disused quarry [616597] above the Gravesend road about ½ mile N.E. of Wrotham church described on p. 107, is succeeded upwards by Upper Chalk (*planus* Zone), of which the following is a section in downward sequence: lumpy chalk with flints partly in bands, 8 ft; band of nodular flints; rather lumpy chalk, 3 ft; band of hard brown nodules, 6 in; rather lumpy chalk, 1 ft; band of semi-tabular and nodular flints; rather lumpy chalk, 6 ft; marl seam, 4 in; soft mealy and lumpy chalk with intensely hard nodular patches containing gastropod moulds, etc., 2 to 3 ft; band of nodular flints, not well developed; lumpy chalk, 2 ft; on a band of large semi-tabular flints.

Osborne White (*in* Bennett 1907, pp. 120, 122) concluded from examples of *Micraster leskei* (Desmoulins) that the horizon is low in the *planus* Zone, but was mistaken in thinking that distinctive nodular 'Chalk Rock' beds might occur at the extreme top of the pit. In fact, the beds wholly comprise lumpy chalk with relatively thin hard nodular bands throughout the zone; only at the base of the zone is the nodular bed below the marl seam rather better developed. At this quarry the basal nodular band has yielded abundant ossicles of *Isocrinus*, gastropod moulds, *Dentalium turoniense* and *Scaphites geinitzi*. This fauna has not been found in any lower nodular bands (see p. 100); its lowest occurrence, with its associated lithology, marks the boundary between the Middle and Upper Chalk which Osborne White assumed to lie within a group of passage beds. Tests of *Micraster praecursor* from the overlying beds show characters typical of the lower part of the *planus* Zone and *Holaster* (*Sternotaxis*) *planus* is abundant. The top 5 ft contain *Gauthieria radiata*, '*Turbo*' *gemmatus* and rhynchonellids in some abundance, including *Cretirhynchia minor* and *Orbirhynchia reedensis*.

In the dry valleys and on the dip-slope of the Chalk tract, *planus* Zone chalk was exposed in a number of places near Tatsfield and Biggin Hill. In the valley running northwards from Botley Hill, at an old pit [39635685] 300 yd N.N.E. of Cheverells Farm, 5 ft of hard rough chalk with flints overlie 5 ft of softer chalk with few flints; 500 yd S. of Bedlestead, on the opposite side of the valley [39935720], somewhat lumpy chalk with bands of large flints has yielded *Dentalium turoniense* and *Limatula wintonensis*. The approximate base of the Upper Chalk is traceable from occasional outcrops of hard yellowish nodular chalk on the bare valley sides. About 450 yd N. of Skid Hill an old pit [40156027] exposed fibrous nodular chalk, with a streaky grey matrix, containing *Micraster praecursor* of *planus* Zone type, sponges and moulds of gastropods. A little farther north, Dewey and others (1924, p. 26) noted streaky grey chalk with fibrous green nodules and moulds of gastropods and other fossils frequently cemented into rock. Approximately ½ mile S.W. of Lusted Hall Farm, formerly Lusted, two overgrown pits [40485716 and 40505693] exposed lumpy yellowish chalk with *Holaster* (*Sternotaxis*) *planus*, *Micraster praecursor* and, in the southerly pit, *Inoceramus costellatus*. From marly chalk with hard siliceous lumps 850 yd S. [40915830] of Norheads specimens of *Micraster praecursor* and *Echinocorys scutata* of typical *planus* Zone type were obtained, while 300 yd farther N. [40995855] flaggy and lumpy chalk towards the top of the zone yielded *Micraster praecursor* showing higher zonal modifications, together with *Holaster* (*Sternotaxis*) *placenta* Agassiz. A small pit [40855950] 400 yd N. of Norheads showed 10 ft of rough chalk of variable hardness with flint bands; the fauna includes *Holaster* (*Sternotaxis*) *planus*, *Micraster praecursor*, *Gibbithyris semiglobosa* and *Spondylus spinosus*, the latter common. On the Biggin Hill side of the valley, chalk of the *planus* Zone was exposed in several old pits. Nodular chalk with sponges and with several species of *Inoceramus* occurs towards the bottom of the valley [41495868] 800 yd S.E. of Norheads, but north of this a local rise of the base of the Upper Chalk to well above the 500 ft contour is indicated by sections in two pits at Biggin Hill. The southern pit [41745910], about 900 yd E. of Norheads, exposed 18 ft of soft chalk with some hard patches and bands of flints. In addition to *Holaster* (*Sternotaxis*) *planus* and *H*. (*S*.) *placenta*, *Cardiaster cotteauanus* d'Orbigny, *Micraster corbovis* and several examples of *Micraster praecursor* were obtained. The last, which were from the middle 4 ft, show characters indicating the upper part of the zone. The base of the zone accordingly is placed at about 520 ft O.D. The northern pit [41505977] ½ mile W.N.W. of Costains Farm, exposed 20 ft of beds low in the *planus* Zone; this indicates the boundary to be at about 540 ft O.D. The fauna from this locality includes the zone fossil, *Micraster praecursor* of low zonal type and '*Turbo*' *gemmatus*.

Massive streaky nodular chalk of the lower part of the zone with the *reussianum* fauna has been noted about 1000 yd N.E. [41006035] of Skid Hill House. Dibley (1918, pp. 81, 92, 93) gives a faunal list which includes characteristic species such as *Metaptychoceras smithi* (Woods) and *Subprionocyclus neptuni* (Geinitz). Noteworthy from this locality in the Geological Survey's collections are *Parasmilia centralis*

(Mantell), *Micraster leskei, M. praecursor* and *Eutrephoceras sublaevigatum* (d'Orbigny). Dewey and others (1924, p. 26) noted here about 20 ft of very fossiliferous nodular rock; the absence of a marl seam led Davis (1926, p. 215) to suppose this had died out, but in fact the rock band of this section probably lies at a higher level. Soft chalk, with hard yellowish patches, from 600 yd W.N.W. [41635964] of Costains Farm yielded *Inoceramus costellatus;* the horizon is placed 40 to 50 ft up in the *planus* Zone. A road cutting [41755860] up the valley side about 1000 yd S.E. of Norheads showed hard nodular chalk with flints in the upper part of the zone, followed by yet harder chalk which lies apparently at the base of the succeeding zone (see p. 113). About 300 yd S. [41305657] of the Old Ship Inn at Tatsfield it is very hard and nodular, with sponges, and is associated with a 6-in marl seam; 400 yd N.E. [41645708] of the inn similar chalk with an 8-in marl seam is probably a continuation of this, and a northerly dip of a little under 3° is thereby indicated. On the opposite side of the valley and 400 yd E. of the last-named site an old pit [42025715] formerly exposed mealy and hard smooth chalk yielding *Micraster praecursor* and the zone-fossil. About 1300 yd S.E. of Norheads hard dense nodular chalk at about 500 ft O.D. [41745837 and 41865828] is referable to the *planus* Zone, as also is rough mealy chalk with flints noted in an old pit [42505691] ½ mile N.W. of Betsom's Hill, from which *Gauthieria radiata* and *Micraster praecursor* of *planus* Zone type were obtained. A small excavation [426578] 650 yd W. of the inn at Westerham Hill exposed beds near the base of the zone; there 6 ft of rough chalk with hard nodules overlie a marly seam, with nodular chalk below. The fauna includes cf. *Sporadoscinia stellifera* (Roemer), *Micraster praecursor,* '*Turbo*' *gemmatus, Inoceramus costellatus* and *Scaphites sp.* Higher beds of the zone, with *M. praecursor,* were exposed in a small pit [431573] 600 yd S.S.W. of the same inn.

In the digitate valley system west of Single Street nodular chalk in a powdery matrix appears in places on the valley sides and hard nodular chalk with a 6-in grey marl seam and yielding *Holaster* (*Sternotaxis*) *planus* was formerly to be seen in a small pit [42455990] 500 yd N.E. of Costains Farm. The level of this horizon, about 580 ft above O.D., is an indication of local minor folding (see below and p. 14), since a section about ¾ mile E.N.E. of Costains Farm showing nodular chalk with a 6-in marl seam is precisely comparable. One effect of this flexure is that northward the valley continues to be floored by Upper Chalk brought down by the northern limb of the fold.

In the broad valley west of Horns Green and Cudham, intensely hard nodular chalk with sponges, the nodules yellowish or grey-hearted, was seen about 600 yd N.W. [44405773] of Bomber's Farm and is near the base of the *planus* Zone. A small pit [44335814] 650 yd S.S.W. of Newbarn exposed 9 ft of rather hard white chalk with flints and some very hard yellowish and grey nodules; this is probably higher in the zone. About 500 yd S.S.E. of Single Street an old pit [43805953] exposed 12 ft of hard nodular chalk, with a 6-in grey marl seam in the middle. Lithological evidence suggests that this may be at the same horizon as the similar sections N.E. and E.N.E. of Costains Farm and comparable with nodular chalk associated with a similar marl seam near Tatsfield, all noted above. Probably the same horizon, which appears to be 30 to 40 ft up in the zone, is indicated in the valley to the east at a point [45606005] 500 yd N.N.E. of Cacket's Farm. This section comprises about 12 ft of nodular chalk, very hard in the upper part, divided by the 6-in seam of grey marl, and another exposure [45466024], on the opposite side of this valley, 250 yd to the north-west, showed hard nodular *reussianum* chalk of the *planus* Zone with *Inoceramus costellatus* and moulds of gastropods. Good sections at both localities were recorded by Dewey and others (1924, p. 27). The relative positions above Ordnance Datum of the sections interpreted as exposing the same marl seam are in agreement with an easterly extension of the minor fold discussed on p. 14. The effect of this roll is to be seen on the out-crops in the Cudham valley: the northerly dip is arrested from Berry's Green north-wards for ⅛ mile, the base-line of the Upper Chalk not falling below the 500 ft contour until nearly opposite Single Street, where it descends comparatively steeply. Its effect in the next valley to the east is to bring Middle Chalk down too low to be exposed in the valley floor.

Farther down the Cudham valley on the east side, a pit [44276056] 200 yd W.S.W. of Hostye Farm, described by Dewey and others (1924, p. 26), showed hard nodular chalk which contains sponges, *Holaster (Sternotaxis) planus, Stereocidaris* cf. *sceptrifera* (Mantell) as well as *Cretirhynchia minor* and ostreids. From another pit [44236077] mentioned 250 yd N., also in the *planus* Zone, were obtained sponges and *Micraster praecursor.*

North of Knockholt an old pit [47235985] 300 yd N.W. of Blueberry Farm exposed 9 ft of mealy chalk with hard nodules and flints and a thin marl seam 3 ft from the top; sponges and moulds of gastropods are here believed to mark the base of the *planus* Zone. Hard chalk, near the bottom of the valley [46926006], about 400 yd N.W. is probably also in the *planus* Zone. The Upper Chalk base-line falls rather rapidly northward on account of the flexure described above.

East of the Darent, Upper Chalk forms the steep sides of the valley near Paine's Farm. In the palmate valley-head to the east of this, *planus* chalk emerges from either side of a broad shallow depression where Clay-with-flints rests directly on Middle Chalk. Hard nodular chalk with gastropod moulds was exposed here and there.

Chalk with flints and yielding *Holaster (Sternotaxis) planus* was noted in an old pit [590606], mostly obscured, at Terry's Lodge (800 yd N.E. of Old Terry's Lodge). Dewey and others (1924, p. 29) referred this to the upper part of the *planus* Zone.

Micraster cortestudinarium Zone. In the region west of the Darent a thin capping of chalk of this zone is present locally. Sections noted are for the most part in the lower beds of the zone.

A roadside pit [39805824] 600 yd N. of Bedlestead exposed 15 ft of rough mealy chalk with harder patches and bands of flints. *Micraster praecursor* occurs in abundance and the evolutional features of the tests indicate a horizon at the base of the zone. *Holaster (Sternotaxis) placenta* was also found. In the narrow valley to the east, yellowish rough chalk with flints from an overgrown pit [40425850] 800 yd S.W. of Norheads yielded *Micraster praecursor* indicative of the upper part of the zone. The relationship of these two exposures suggests that hereabouts there is a local swing of the strike to a south-westerly direction. Chalk of this zone fringes the edge of the Clay-with-flints near Norheads, and up the valley towards Lusted Hall Farm, and there may be a little on the west side of the main valley towards Skid Hill. The only section noted was in an old pit [40785767] 300 yd W. of Lusted Hall Farm, where hard white and smooth yellowish chalk with flint bands yielded *Micraster praecursor* of low zonal type, a large gibbous variety of *Echinocorys scutata* and *Inoceramus inconstans striatus* Mantell. Owing to the dip, though chalk of the zone does not appear in the upper reaches of the valley system north of Tatsfield, it may well fringe the Clay-with-flints plateau farther north towards Pimlico.

In the steep road cutting [41765860] about 1000 yd S.E. of Norheads, intensely hard nodular chalk at about 580 ft O.D. may represent the 'Top Rock' horizon noted in the adjacent Reigate and Dartford areas. A little to the north of this point, towards Biggin Hill, the zonal outcrop is cut off by the gentle transverse fold described above, and it probably reappears on the northern limb of the flexure only just within the boundary of the map area on the hillside north of Saltbox.

In the valleys near Single Street, Cudham and Knockholt there is similarly little room for chalk of this zone. A section at the head of the dry valley north of Knockholt Pound, however, indicates a capping on the higher ground. The pit [48256071] is situated in Deerleap Wood, 250 yd N.E. of Park Farm, and was recorded by Dewey and others (1924, pp. 30, 40). The chalk is white and massive and contains the zone fossil *Echinocorys scutata* and *Micraster praecursor.*   S.C.A.H.

Chapter VI

PLEISTOCENE AND RECENT

GENERAL ACCOUNT

THE MOST widespread superficial deposits of the present area comprise large sheets of heterogeneous unbedded material on the Chalk and the Lower Greensand, and less important spreads of similar composition on the Weald Clay outcrop. Probably all of these are more or less related and many of them reflect periglacial solifluction, although those on the Chalk owe their lack of structure in some degree to solution of chalk beneath them which has produced disruption of any bedding that may have once existed. These last come mainly within the Clay-with-flints formation but also included are restricted areas of flint pebble beds originally belonging to the Blackheath Beds of Tertiary age. These are distinguished from the Clay-with-flints by being named Disturbed Blackheath Beds.

The remainder of the deposits are grouped under the common name Head, a name first given to a mantle of scree-like material on the hill slopes of Devon and Cornwall. These drifts have been subdivided, partly on their general location, and partly on their constitution, into Angular Chert Drift, Limpsfield Gravel, Scarp Drift and Downwash, Head Deposits on Weald Clay, Head Deposits on Hastings Beds, Dry Valley Gravels and Associated Deposits, and Chalk Breccia; they are described below under those headings. Though Dry Valley Gravel is indicated separately on the One-inch map, the other deposits are either shown as Head or, in the case of Scarp Drift and Downwash and of Chalk Breccia, have been omitted from the map on account of their local and impersistent nature. Bedded Drift deposits, associated with river development, include terraced river gravels and brickearths of Pleistocene age and alluvium of Recent age. On the whole, gravels are relatively unimportant. Nowhere do they occur in extensive spreads, and for the most part they are but a few feet thick, although thicknesses of 10 to 20 ft or more have been proved locally. Similarly, brickearths are unimportant, except for extensive deposits in the Medway valley, around Hadlow, East Peckham, Paddock Wood and east and west of Tonbridge.   F.H.E.

DISTURBED BLACKHEATH BEDS

Five areas of Disturbed Blackheath Beds have been noted, all in the neighbourhood of Knockholt and Knockholt Pound. Of these, one [477584] about ¾ mile E.S.E. of Knockholt church lies on the Chalk outcrop and almost isolated from other Drift deposits and two, at Shelleys [463593] and a little east of Knockholt [472588] respectively, are partly on Chalk and are partly associated with Clay-with-flints. The other two, about ½ mile E. and W. respectively of Knockholt Pound [490598 and 475595], occur wholly within the Clay-with-flints deposits of the Knockholt area.

In addition to these, Blackheath Beds material is present in much of the Clay-with-flints, with which it is so closely incorporated as to be incapable of separation. S.C.A.H.

CLAY-WITH-FLINTS

Clay-with-flints is a mixture of stiff brown and reddish clay with large unworn flints, together with subsidiary sand, loam, pebbles of Tertiary age, sarsens, chert and ironstone from the Lower Greensand, and ironstone of probable Pliocene age; patches of deeply patinated reddish brown flints occur here and there. It is confined to the Chalk plateau area and normally caps the high ground only. A certain amount of downwash, mantling the heads of the dry valleys and the upper slopes of the valley sides is not uncommon, and occasionally slipped masses of Clay-with-flints spread over parts of the upper slopes of the Chalk escarpment.

Sands derived from the Tertiary Beds are bright in colour, often vermilion and orange. Owing to solution of chalk by percolating water the Chalk surface below Clay-with-flints is very irregular, and accordingly the thickness of the drift is exceedingly variable from place to place. The average thickness of the deposit is rarely above 6 ft, but drift-filled solution pipes in the Chalk are commonly 50 ft or more deep. The base of the deposit is often stained with manganese, which imparts a characteristic purplish black coating to the enclosed flints. A section exposing Clay-with-flints near Westerham is shown in Plate VIIIA.

More than a hundred years ago Trimmer (1852, p. 275) noted on the edge of the scarp at Wrotham Hill, north-west of Wrotham, a difference between Clay-with-flints *in situ* and the calcareous downwash covering the hill slope. Prestwich (1855, pp. 70, 71, 73) described pipes in the Chalk above Westerham and Wrotham, and Chandler and Leach (1909, p. 237) refer to pipes exposed in a Chalk pit ('Shore Hill Quarry', 549594) a little over $\frac{1}{2}$ mile N.W. of Kemsing church, later described in detail by Leach (1921, p. 38) with special reference to sands of Tertiary origin there associated with the Clay-with-flints.

Pipes exposed in 1930 in an old pit [55025955] at a higher level than the above about $\frac{1}{2}$ mile E. of Otford Court, formerly Beechy Lees, showed Clay-with-flints to extend as a continuous layer over the top of sand-filled pipes, themselves lined with Clay-with-flints. Considerable attention has been given to problems relating to the Tertiary sands associated with the Clay-with-flints in the area. Prestwich (1858, p. 324) classed sands above Otford with the Pliocene, as did Lewis Abbott (1916, pp. 183, 184, 188) in referring to sandstones at Otford Mount [538597] ($\frac{1}{2}$ mile E.N.E. of Otford Station) and above Kemsing, and sand overlying conglomerate north of Wrotham. Whitaker (1872, p. 340) recorded local masses of sand of uncertain age. Plastic crimson-mottled clay, similar to that of the Reading Beds, is also fairly widespread over the Clay-with-flints area. Many small masses of pebbles are present in the area north of Tatsfield and eastward to Knockholt. Wooldridge (1927, p. 61) has suggested a Blackheath or earlier age for them and considered them to be distinct from Pliocene deposits. Ironstone of a type different from that occurring in, or derived from, the Lower Greensand is, however, associated with the sand and Tertiary pebbles about 300 yd N. of Shelleys [463593], near Knockholt. The late Professor Wooldridge also considered the sandy beds above Otford to be

J

of Eocene age; he mentions that sand piped into the Chalk at Wrotham Hill however, has a distinctive heavy mineral assemblage, indicating southerly overlap of the basal shingle by sand laid down by the Pliocene sea in this area (Wooldridge 1927, p. 63).

Details of a temporary section near Saltbox [411604] (now demolished but formerly 1600 yd N.W. of Costains Farm), north-west of Biggin Hill, given by Berdinner (1936, p. 15) demonstrate the complex and disturbed nature of the Clay-with-flints and the associated sands and pebbles. Desilicified flints occur in pipes at Wrotham Hill, over ½ mile N.W. of Wrotham church. Comparable, but more extensive, desilicification phenomena have been described in a disturbed outlier of Eocene beds, just beyond the boundary of the map area at Knockmill [575614], 1¼ miles N. of St. Clere (Chandler and Leach 1936, p. 239).

Patches of a type of drift, probably of southerly derivation, are locally associated with the Clay-with-flints on the Chalk plateau. Prestwich (1890, p. 157; 1891, pp. 127, 129) noted numerous thin sprinklings of old flint and chert drift on the hills above Wrotham and Kemsing and above Chevening and Titsey. These gravels include cherts derived from both Hythe and Folkestone Beds as well as Lower Greensand ironstone and ragstone. Prestwich was indebted to Benjamin Harrison, De Barri Crawshay and others for their observations on many localities where flints thought to be human artifacts had been found. Harrison's work on the plateau drifts in the area was dealt with in some detail by Bennett (1907, pp. 12–14, etc.) who described the beds seen in pits which were sunk near (Old) Terry's Lodge [587600]. Bennett and Harrison (1906, p. 463) record the presence in them of derived Gault fossils. Bury (1910, p. 648, etc.) realized the significance of southerly derived materials in these drifts but was unable to agree with Prestwich that their transport was subaerial. Salter (1905, p. 7) included in his summary of high level drifts localities west of Otford and at Knockholt Beeches where he saw deposits containing Tertiary debris. The late Sir Edward Harrison (1928) assembled numerous records of Benjamin Harrison's observations on plateau drifts in the district, and more recently (1958) assessed the validity of his father's study of supposed eoliths in relation to these drifts and their probable age.

S.C.A.H.

HEAD DEPOSITS

ANGULAR CHERT DRIFT

Angular Chert Drift occurs on the Lower Greensand outcrop and is similar in mode of occurrence to the Clay-with-flints of the Chalk Downs. It consists of a structureless mass of angular chert fragments set in a matrix of loamy sand, often bright red in colour, due probably to oxidation of glauconite derived from the Sandgate Beds or other beds of the Lower Greensand. This drift is extensively distributed over the dip-slope of the Lower Greensand and generally supports dense woodland, largely of chestnut. It was extensively worked in the past, and its spreads are usually covered with small pits, up to 4 ft deep, mostly overgrown. From the escarpment crest, where the deposit rests upon the upper part of the Hythe Beds, it sweeps down the dip-slope and overlaps the Sandgate Beds. In the neighbourhood of Ightham, the development of chert layers in a glauconitic loamy matrix occurring near the top of the Folkestone Beds

A. Heathland on the outcrop of Folkestone Beds: Oldbury Hill
in the distance

(A 7113)

PLATE IX

B. Palaeolithic Rock Shelters, Oldbury Hill

(A 10377)

(p. 60) furnishes material for angular chert drift at Raspit Hill and Oldbury Camp. Bennett (1907, p. 18) classified the chert drift as 'Residual Drift' and Prestwich (1889, pp. 275–6) has described its occurrences near Offham and Ightham. Like the Clay-with-flints, the Angular Chert Drift extends in general down to about 450 ft O.D. and is dissected by the dry valleys and coombes of the dip-slope. Being one of the older drifts of the area (Dines, Hollingworth and others 1940, p. 220) it has contributed to a large extent to the later head deposits on the Weald Clay and in Holmsdale (p. 121). A typical section on Limpsfield Common is shown in Plate VIIA. In the neighbourhood of Offham, the extensive deposits of Mereworth Woods extend from about 550 ft above O.D. to below 300 ft and their lower-lying portions contain scattered ochreous flints, presumably incorporated from a pre-existing drift of flint gravel. This lower part may therefore be of later date than that nearer the escarpment crest. Flints are also to be found sparsely scattered over the lower parts of the areas of Angular Chert Drift south of Brasted and Sundridge. Some very small outlying patches of this drift were grouped as higher terrace gravels by Topley (1875, pp. 191–3).

On the crest of the Lower Greensand escarpment the drift rests directly on thick chert layers *in situ* in the top of the Hythe Beds, drift and solid being worked together for the stony material. With the thinning out or disappearance of the chert bands of the Hythe Beds when traced northwards down the dip (see p. 57), the rubble drift alone contains stone, and as a consequence most of the shallow workings on the lower parts of the dip-slope have long since been abandoned as uneconomic; the few active or recently active workings are situated close to the escarpment crest.

Sections vary little from place to place. Generally the deposit supports podsolic soils with a dark peaty layer of leaf litter at surface, below which lies from 1 to 1½ ft of white or pale grey sandy loam with bleached sand grains, containing angular chert fragments which are themselves patinated white. The bottom few inches of this layer are usually a much darker grey than the remainder, sometimes approaching a lilac colour. Below the soil proper comes the drift, comprising reddish brown to buff mottled loamy sand, resembling a brickearth, with occasional small rafts of green glauconitic sand, many angular chert fragments, generally pale brown or honey-yellow, and occasional fragments of ferruginous sandstone. The deposit frequently shows frost contortion and festooning.   H.G.D.

LIMPSFIELD GRAVEL

An interesting deposit of gravel which lies east of Limpsfield, on the watershed between the Darent and a Medway tributary, was described by Topley (1875, pp. 193–4), as a 'watershed' gravel; he was of the opinion that it could not be referred to the Darent. Other patches of similar material, recorded as far east as Westerham in the Darent Valley, he regarded as watershed gravel re-sorted by the local streams when they were more important than they are now.

Prestwich (1891, pp. 137–44) described the Limpsfield gravel and similar patches extending eastward of Westerham and regarded them as river deposits which might be correlated with high-level gravels of the Thames. Treacher (1909, pp. 60–1) too, believed the gravel to be part of a terrace of the Darent, and he accounted for its present position by the capture of the head waters of an early Darent by the Oxted stream, a tributary of the Medway. Gossling (1941a) concluded that the Upper Darent Valley contained gravels of two types: solifluction deposits and terraced river gravels.

Lack of bedding but the presence of a stiff clay matrix with unworn flints in the Limpsfield deposit are points against its being a river gravel, although in the vicinity there are traces of a drift which in all probability formerly belonged to a river terrace (see p. 128). The nature of the constituents suggests that the deposit is an accumulation of debris, brought mainly from the north, but with contributions from the south, on the floor of an east–west valley of which the patches of high level drift to the east are also the remains.

In 1934 a trench dug across the Limpsfield spread from north to south immediately east of Limpsfield showed that the most abundant constituents are blue flint pebbles, of all sizes up to 6 inches in long axis; some are broken and have received subsequent wear. Next in abundance are angular and sub-angular fragments and nodules of pale blue flints and angular fragments of yellow flints. Battered brown flints, pieces of conglomerate, occasional pale grey quartzite pebbles, and small chips of soft brown shale are also present. In the southern part of the deposit there are, in addition, angular and rounded pieces of ironstone and siltstone and, in the extreme south, subangular fragments of chert. The flint pebbles and also the conglomerate can be matched in patches of Disturbed Blackheath Beds which lie on the dip slope of the Chalk to the north; e.g. at Worms Heath [380580], some 3 miles N.N.W. The pale blue flints have been derived from the Chalk, or from the Clay-with-flints in which also yellow flints and battered brown flints are to be found. Quartzite similar to that of the pebbles occurs in the Folkestone Beds at Sundridge and may occur also in the Limpsfield area; the ironstone resembles the carstone of the Folkestone Beds from which also the siltstone may be derived. Chips of shale probably come from the Gault, and the chert from the Hythe Beds. Gossling claimed a Wealden origin for pebbles of siltstone and ferruginous mudstone which occur locally.

Although the deposit is mainly a structureless mass with the constituents unevenly distributed, locally it is crudely bedded. About the centre of the patch there was noted a group of pebbles orientated with their long axis vertical and their flat side to the north. At the northern end, where the deposit overlies the Gault, the stones are embedded in a stiff grey or brown clay matrix, weathered buff or yellow near the surface, and occasionally lenses of clay without stones occur within the main mass. Near the northern extremity of the deposit, wedges of gravel with their thin ends towards the north occur between layers of clay resembling Gault; both clay and gravel contain seams of sand.

The clay persists as a matrix south of the margin of the Gault for some distance over the Folkestone Beds, beyond which the matrix passes through a loamy zone into a sand similar to that of the Folkestone Beds; this continues over the Sandgate Beds.

On the north and west the surface of the drift is over 500 ft above Ordnance Datum (maximum 520) and it slopes gently to a level of about 476 ft on the south-east. The maximum thickness observed was 10 ft. The Limpsfield gravel was once extensively worked, but the many pits are now overgrown.

Prestwich recorded the finding of pointed and ovoid Palaeolithic implements, including Acheulian types. Implements were also noted by Treacher.   S.B.

PETROGRAPHY OF PEBBLES IN THE LIMPSFIELD GRAVEL

Selected specimens examined, in thin section and by heavy mineral concentrations, came from east of Limpsfield, on the northern side of Limpsfield Common. The pebbles are siltstones of buff to brown colour, measuring up to about ½ inch in length and are of angular to flaky shape (British Standard 812: 1960, p. 16). They are composed of angular quartz grains, mainly 0·006–0·015 mm across, with flakes of muscovite and chlorite, films of clay minerals and granules of iron ore and other heavy minerals reaching 0·1 mm in size (E 25598).

Separation of the heavy minerals yielded a distinctive suite, but the amount of material available did not allow precise figures to be determined for the number of grains present of each species. The heavy minerals constitute appreciably less than 1 per cent by weight of the sand. A few pebbles were crushed, boiled to remove iron oxide with 1 : 1 hydrochloric acid in the presence of excess stannous chloride and separated in bromoform of specific gravity 2·89. The heavy mineral crop was found to consist of abundant yellow sphalerite, common opaque mineral grains, common zircons forming both elongate idiomorphic prisms and rounded grains, scarce tourmaline in brown, green, blue and rarely in parti-coloured green and brown grains, scarce andalusite, epidote, pinkish garnet and rutile grains and rare anatase, baryte and brookite. To confirm the identity of the sphalerite, the presence of zinc was proved by microchemical testing, in collaboration with C. O. Harvey, using as reagents copper sulphate and ammonium mercuric thiocyanate in a solution slightly acidified with sulphuric acid.

Two features of the very distinctive heavy mineral suite call for comment: the abundance of sphalerite and presence of rare baryte, and the scarcity of rutile. Sphalerite appears only to have been recorded in the sediments of the south-east of England from the fuller's earth and associated sands of the Sandgate Beds of Nutfield, Redhill and Tilburstow Hill (Davies 1916, Newton 1937), all of which lie west of the superficial gravel now under investigation. Baryte has been recorded in southern England from various horizons and localities (e.g. Davies 1916) including the fuller's earth, where sphalerite and baryte tend to occur in association. Both are present in all three samples of this rock-type among the nine samples of Sandgate Beds examined by Davies, and baryte was noted in two of the other Sandgate Beds samples which he examined. Among the remaining 62 samples from other formations of the Croydon 'Regional Survey Area', for which Davies tabulated the mineralogy, only one had baryte present and none sphalerite. Rutile was found by Davies to be "an abundant constituent in nearly all the heavy residues" which he examined but was absent in the fuller's earth. Newton (1937, p. 185) noted sphalerite as common or uncommon in four samples of fuller's earth and as very common in a section of the overlying beds at Cockley Pit, east of Redhill. He commented "zinc blende is present throughout the [fuller's] earth-bearing series, but rare outside these rocks" and noted also that "rutile is rare throughout the earth-bearing series".

Since sphalerite has not been recorded from the sandstones of the Weald or the Tertiary deposits of the London Basin, it seems probable that the mineral has come from the fuller's earth beds, which form a lenticular outcrop in the Sandgate Beds, with its maximum development at Nutfield, and do not extend beyond Oxted (Dines and Edmunds 1933, pp. 58–65). Sphalerite has only been recorded from Redhill to Tilburstow Hill and it is not certain whether it occurs farther east, but in view of the change of facies and attenuation eastwards of the fuller's earth beds it seems improbable that any is present near Limpsfield.

P.A.S.

It seems likely that the pebbles examined came from a general westerly direction along the Vale of Holmsdale, though this westerly source for the Limpsfield gravel is at variance with the views of Gossling (1936, 1937, 1941) upon the drainage of the Vale of Holmsdale. He claimed that pebbles of silt-stone and ferruginous mudstone found at Limpsfield, Westwood and Westerham are of Wealden derivation, a view which he recognized as conflicting with the observations of Topley (1875, pp. 188, 295), who had stated emphatically that the Darent deposits contain no pebbles of Wealden sandstone and had recorded none from Limpsfield.

S.C.A.H.

Scarp Drift and Downwash

Head deposits on the Chalk escarpment and on the more northerly slopes of the Lower Greensand, mostly too sporadic and thin to be shown on the geological map, occupy considerable areas in the Vale of Holmsdale but nowhere reach the extent and thickness of the similar late Glacial and post-Glacial deposits recognized in more recent years in the lower Medway valley, the Isle of Thanet and near Ashford, Kent (see Kerney 1965, p. 274). In some places deposits are well defined but in others they are only small accumulations of downwash or remnants of former spreads. Bennett (*in* Bennett and Harrison 1906, p. 460; Bennett 1907, pp. 14–18; 1908b, p. lxv and pl. II), following Prestwich (1889, p. 270 and pl. IX; 1891, p. 126 and pl. VII) who himself was largely indebted to the observations of Benjamin Harrison, arrived at a classi-fication mainly into (a) Southern Drift (with materials of southern derivation) and (b) a younger Scarp Drift representing the waste from the denudation of the Chalk escarpment. Prestwich described particularly the Southern Drift on the Chalk plateau (see p. 116), which contains Lower Greensand materials and alleged Eolithic implements, but Bennett's Scarp and Southern Drifts include also High Level or Limpsfield Gravel (p. 117), High-level and Lower-level Brickearth and Lower-level Gravel. The patches of drift with materials derived from the north (scarp), south, or locally from the Shode valley ('Old Shode Gravels'), are shown diagrammatically on a map published by Bennett in 1908, who fixed a definite northern boundary to the Scarp Drift but could not strictly define a southern boundary owing to the overlapping and mixing of Scarp and Southern Drift. Patches of drift with northern materials are shown far southwards at Ivy Hatch and Old Soar. Like the chert drift on the Lower Greensand (see p. 117), Bennett termed the chalky scarp drift 'Residual Drift', implying absence of transport in its formation. Produced from wastage of the Chalk escarpment, it has probably been subjected to considerable sludging or downwash movement, perhaps assisted by cultivation, a process which may be observed today after heavy rain or frost conditions and which has caused considerable deposits of chalky drift to accumulate even in historic time.

The passage of chalky drift into loams, e.g. at Borough Green (see p. 131) is ascribed to decalcification. It should be mentioned that the 'Shode Gravels' do not appear to have been river deposited. On the other hand, some of the drifts shown by Prestwich on his Map of the Darent Basin fall naturally into categories such as terraced gravels of the Darent near Dunton Green and Otford, though their upper layers may show signs of later disturbance and rearrangement.

Topley (1875, p. 140) records chalk in sand, the latter probably rearranged Folkestone Beds material, by the roadside going north from Seal towards Otford Mount; small pieces of chalk occur in loamy brickearth at Pascall's Brick and Tile Works north of Platt, near Borough Green.

Head formed of downwashed Clay-with-flints material, locally quite thickly accumulated, occurs here and there on the upper slopes of the Chalk escarpment and to a lesser degree in the Chalk valleys. S.C.A.H.

HEAD DEPOSITS ON WEALD CLAY

Head deposits on the Weald Clay are divisible into three groups on the twin basis of composition and geographical distribution (Dines and others, 1940 p. 221). In the first group, well-rolled, blue flint pebbles are conspicuously abundant and associated with subangular brown flints, chert, buff glauconitic sandstone, and fragments of ironstone. The pebbles and flints are similar to those found in the Limpsfield Gravel, the chert and sandstone are undoubtedly from the Hythe Beds, and the ironstone closely resembles the carstone of the Folkestone Beds. Patches of unbedded pebbly drift occur below a gap in the Lower Greensand escarpment at Limpsfield and extend as far as the Medway, but are confined to the area of the Kent Brook except for a few patches which are to be found at higher levels than the terraces of the Eden at Skeynes [434461], Hever Lodge [477463], and west of Chested [502463]. Several of these patches lie along the sides of the valley of the stream which joins the Eden Brook south of Haxted and on the western slope of the Kent Brook valley, as well as on the watershed between the two streams. These isolated patches of drift are, with one or two exceptions, too thin to have been worked, and exposures are rare.

The second group is characteristically free from flints or pebbles, and consists of subangular and occasionally angular chert, sandstone and sand, all from the Hythe Beds. This chert drift extends as isolated masses from the base of the Lower Greensand escarpment over the Weald Clay as far as the railway between Edenbridge and Tonbridge, the lateral boundaries being the Kent Brook on the west and a line joining Shipbourne and Tonbridge on the east. In most instances it lies on the watersheds of minor streams draining to the Medway and descends to points below the lowest level of the first group, to reach the level of the second terrace of the Medway. The oldest of the three stages of solifluction drift recognized by Weeks (in press) and referred to on p. 144 forms part of this group.

The third division includes patches of subangular battered and pitted flints, cherts, sandstones and occasional well-rounded flint pebbles. Similar flints are to be found in the Vale of Holmsdale where it is also possible to match the pebbles. Chert and sandstone similar to that in the drift occur in the Hythe Beds. This drift is found in the vicinity of the Shode Valley. A rounding of its constituents becomes more apparent as the deposit is traced from north to south, in which direction also it is crudely bedded as far as Hadlow, beyond which it merges into the well-bedded river gravels.

Some of the deposits recorded by Prestwich (1889, p. 270 and pl. IX) are included in these groups; also some noted by Topley (1875, pp. 185–6).   s.b.

The Head deposits of the Weald Clay in west Kent were divided into two groups by Bird (1963, p. 448) corresponding to the first two divisions recognized above. He thought it possible that the flint-bearing gravels of the first group between Oxted and Edenbridge were transported by streams rather than by solifluction. Three terrace features were recognized by him, a low terrace at 120 to 130 ft, a middle terrace at 180 to 200 ft and a high terrace at 300 to 325 ft. The Head deposits extend across the high and middle terrace but not on to the low terrace or the present valley floor. Bird concluded that the formation of the Head must have taken place when the middle terrace was still undissected before the low terrace and present valley floor came into existence. There may have been earlier periods of solifluction, since destroyed by denudation, or later ones which did not give rise to such extensive spreads of chert rubble. The middle terrace is tentatively correlated with the Boyn Hill stage in the Thames valley. Worssam (1964, p. 574) recognized similar, but smaller, patches of Head in the Maidstone district and thought that the formation of the Head, and cambering of the Hythe Beds, had taken place during the Gipping Glaciation.   c.r.b.

Head Deposits on Hastings Beds (Hillwash Head)

Certain valleys south of the Medway and in the eastern half of the map area are partially infilled with downwashed sands. Most of the material is derived from the Tunbridge Wells Sand, or, in more restricted examples, the Ashdown Beds. Where the Tunbridge Wells Sand provides the constituents of the Hillwash Head, the upper (southern) limit of the Head is the spring line issuing from the Tunbridge Wells Sand/Wadhurst Clay junction. The Head extends down the valleys and on to the flood plain of the Medway where it merges with the Brickearth. A similar relationship of Head 'feeding' the Brickearth has been noted in the Chatham area (Dines and others 1954, p. 111).   c.r.b.

Dry Valley Gravels and Associated Deposits

Dry Valley Gravels occur in certain valleys which dissect the Chalk plateau west of Biggin Hill and north of Knockholt. Their composition varies from place to place according to the nature of the beds on the surrounding high ground that furnish the parent material. Normally a stony loam with large irregular flints, slightly battered but seldom showing signs of much rolling, they may consist largely of dark brown clay with smaller angular flints and occasional Tertiary pebbles where Clay-with-flints is present on the higher parts of adjacent valley sides above steep Chalk slopes. Fragments of chalk, often comminuted, occur in the loam, especially near extensive bare chalky slopes. In some places the deposit consists of coarse angular flint gravel with little matrix apart from broken-up flint.

These deposits, continuous with similar gravels of the Dartford area, occur in the deeper and narrower valleys, to which they impart a characteristic flat bottom; they have not been found in the broader valleys, but it is probable that irregular downwash deposits, and certainly some variable chalky rubble and loam, occur in all the dry valleys. Dry valley drifts do not appear to have been water-deposited as river gravels.

A temporary section in these gravels about ¼ mile N.N.W. of Lusted Hall Farm, formerly Lusted, showed clay with angular flints and chalk fragments to be at least 6 ft thick. This is the most southerly point to which continuous drift has been proved; in the higher reaches of the valleys downwash loams and flinty rubble are sporadic, thin, and unimportant, except that they give rise to a soil somewhat different from a normal chalk soil. In the valley ½ mile N.N.E. of Skid Hill House the Jewels Wood borehole (p. 149), sited very slightly above the flat valley floor, passed through 9 ft of chalky rubble with flints. S.C.A.H.

Chalk Breccia

A hard breccia has been noted at a number of scattered localities on the edge of the Chalk escarpment and high up on the bare Chalk valley slopes. It consists of angular, or occasionally rounded, fragments of chalk, with or without flints, cemented in a buff or grey matrix. Both fragments and matrix show all stages from a completely calcareous to completely siliceous rock. The deposit, wherever noted, forms tabular masses of limited lateral extent, each of which appears to mantle the hill slope for a few yards only. No sheet is usually more than 2 ft thick and may be only of the order of 6 in. The breccia is intensely hard, especially when highly silicified.

Breccia has been noted 500 yd S.S.E. of Norheads [41055870], near Biggin Hill (580 ft O.D.); 100 yd N. of Otford Court, formerly Beechy Lees [54255975], near Otford (595 ft O.D.); a little over ½ mile N.W. of Kemsing church (about 600 ft O.D.), as a mass embedded in talus [54915942]; just over ½ mile N.W. of Newhouse Farm [59305985], near Wrotham (710 ft O.D.); a little over ½ mile N. by W. of Newhouse Farm [59686005], near Wrotham (700 ft O.D.), an occurrence noted by Bennett and Harrison (1906, p. 464), Bennett (1907, p. 95) and by Holmes (1937, p. 353); and about ⅝ mile N.E. of Wrotham church (680 ft O.D.). It occurs locally in the adjacent Dartford and Chatham areas (Prestwich 1891, p. 136; Hood 1884, p. 62). With but few exceptions it lies not more than 50 ft, and usually about 25 ft, below the Clay-with-flints capping the Chalk plateau. Chandler and Leach (1932, p. 288; 1936, p. 246) have recorded fragments of silicified chalk from the Clay-with-flints at Knockmill, slightly north of the present area and about 2¼ miles N.E. of Kemsing. In this locality striking deposits of de-silicified Tertiary pebbles have also been described and it has been suggested that there may be some connexion between this and silicification of local chalk. There may be a similar relation between the silicified chalk breccia of the Sevenoaks area and its proximity to Clay-with-flints and its associated Tertiary materials. S.C.A.H.

River Gravels, Brickearth and Alluvium

The River Deposits of the area are associated with the middle reaches of the River Medway, its tributary the River Eden, and the headwaters of the River Darent.

Terraced River Gravels

In both river valleys gravels of four terraces have been recognized; all four have been much dissected by erosion. Sections in all but the lowest of these were described by Topley (1875, pp. 184–93). The Fourth, and highest, terrace of the Medway is represented in the present map-area by two spreads, lying about 80 or 90 ft above the river, one about 2 miles N.N.E. of Tonbridge, the other 1 mile N.N.E. of Hadlow. In addition there are two very small patches about the same level at Nettlestead Green and one ¾ mile E. of Capel. In the Darent valley seven small spreads have been noted at about 120 to 150 ft above present river level, all in the neighbourhood of Dunton Green, Sevenoaks and Seal. Gossling (1941a) distinguished both river gravels and solifluction gravels, however, between Limpsfield and Westerham; but as the former are usually concealed by the latter it is not possible to separate high terrace gravels where exposures are absent.

The Third Terrace of each valley is equally fragmentary. In the Medway valley it has been preserved in one spread to the north of Tonbridge, where it underlies part of the town, and another at Meopham Bank, formerly Meopham Park, south of Hildenborough; it is usually between 50 and 60 ft above river level. In the Darent valley there are two small spreads westwards of Brasted, two near Dunton Green and one near Seal, but gravel described by Gossling at Westwood, west of Westerham, where it is overlain by head of the Limpsfield Common type, may be referable to this terrace. From a section which Prestwich (1891, p. 147) described in the deposit south of Dunton Green it appears that this gravel has been much disturbed by solifluction or may be wholly of head origin.

The gravels of both terraces, in both valleys, consist mainly of flints and debris from the Lower Greensand, with a certain amount of Wealden material in the Medway terraces, although in general this is too soft to have been preserved as pebbles. The ratio of sand and loam to stone is high. The gravels are not usually thick but 10 ft were proved at Tonbridge and 18 ft at Seal.

In both valleys the Second Terrace predominates and in each case dissection by erosion is marked; but few of the various spreads, all lying about 30 ft above river level, exceed half-a-mile either in width or in length. In the Darent valley, however, a thickness of 17 ft was proved in a trial borehole at Riverhead. As in the case of the higher terraces, the ratio of stone to sandy matrix is low, and nowhere in either valley is gravel of this terrace quarried today.

In the Medway and Eden valleys gravel of the First Terrace occurs at the western and eastern edges of the map. Thin deposits, lying a few feet above the river, border the alluvial belt near Holland and about two miles south of that place; others extend around Edenbridge, particularly on the west. Gravel is present on the south-east side of the river at Hale Street, near East Peckham. The First Terrace in the Darent valley grades into deposits, buried under or mixed with the Alluvium, which have been proved to a depth from surface of as much as 38 ft in one place at Sundridge, 12 ft or more at Riverhead and 20 ft between Riverhead and Otford. A small isolated patch less than ¼ mile N.E. of Westerham church has been referred to this terrace, but the small spreads near Dunton Green and between Riverhead and Sevenoaks are bordered directly by Alluvium.   S.C.A.H.

Buried Channel Deposits

In the alluvial belts of both the Medway and Darent river systems deposits of rough gravel are of local occurrence beneath overlying silts, peat and mud. Commonly they are at shallow depth and are contiguous with First Terrace deposits, but they may also occur as irregular buried stream-courses filled with the coarser drift. The overlying Alluvium was laid down during a subsequent period of aggradation, when the previously incised valleys became 'drowned', and not only did the gravels accumulate variably, according to the conditions of stream-flow, erosion and available pre-existing sedimentary materials, but also much of any finer, sandy, content the gravels retained was subject to removal by movement of water within them, rendering them poorly consolidated.   S.C.A.H.

Brickearth

Brown or buff, fine, more or less structureless loamy drift of this type is comparatively thinly developed in the Sevenoaks district, and the greater part is assigned to 'River-Brickearth', as certain deposits in the Chatham area have been classed by Dines, Holmes and Robbie (1954, p. 112).

Of the brickearths additional to the mixed gravelly and loamy deposits described with Scarp Drift and Downwash, the most important and widespread occurrence is in the Medway Valley around Tonbridge, Hadlow, East Peckham and Paddock Wood. The material there consists largely of loamy silt, with an admixture of sand and locally some gravel; it probably accumulated as a result of ponding of the Medway above the river gap through the Lower Greensand ridge at Yalding, a little eastward of the present area.

Some of the small patches elsewhere, as near Oldbury Hill and West Malling, may have originated as Head deposits.   S.B.

Alluvium

The most extensive area of Alluvium is in the Medway valley above and below Tonbridge; the well-defined belt reaches a mile in width some three miles east of the town. In the Darent valley it reaches only $\frac{1}{2}$ mile in width, between Riverhead and Otford. These are also tracts in which thicknesses up to 13 ft are recorded, the amount elsewhere being frequently less than 5 ft.

The deposits are mainly soft clay, mud and silt, with a variable content of impure sand and stony loams. Peat occurs locally, $3\frac{1}{2}$ ft being recorded at Riverhead. The spreads on the Gault and Folkestone Beds about 1 mile S.E. of Kemsing and northward of Ightham are largely local downwash of variable flinty or sandy clay and loam, partly gravelly in the lower part, and in boreholes proved up to about 10 ft thick. The deposit near Ryarsh is similar.

Eastward of Ryarsh the Alluvium is dark, clayey and peaty, with but few flints. A section [64405856] in Addington Park 140 yd E.S.E. of St. Vincents showed 5 ft of dark brown mottled loam with occasional flints and fragments of ironstone and chert.   S.C.A.H.

Landslip

Landslips occur on the face of the Hythe Beds and to a lesser extent on the Lower Tunbridge Wells Sand escarpment. In all cases it is the clay immediately below the junction with the overlying sand which has moved, a direct result of weakened, water-saturated clay below the sand/clay spring line. The effect of these slips on topography is the production of hummocky, poorly drained ground and locally, trees showing adjustment to verticality. Although slipping initially commenced during the warmer conditions following the Ice Age there is evidence for continuing instability to the present day. Topley (1875, p. 316) quotes slips with up to 65 ft displacement and involving 9 acres of ground which took place at Westerham in 1596, and also one at Toy's Hill in 1756.

C.R.B.

DETAILS

DISTURBED BLACKHEATH BEDS

Pebbly sands with ironstone occur in the area north of Shelleys. Brown loamy sand with pebbles and subangular flints overlies coarse-grained orange and brown sand about ¼ mile E.S.E. of Knockholt church [47175884], while in the wood immediately to the south-east pebbly soil and orange-coloured sand are present. Sand with many pebbles occurs on the Chalk spur [477584] ¾ mile E.S.E. of Knockholt church. Abundant flint pebbles and orange loamy sand with some subangular flints mark the areas near Knockholt Pound.

S.C.A.H.

CLAY-WITH-FLINTS

Many Tertiary pebbles occur south-west of Skid Hill, and 300 yd S.S.W. [408588] of Norheads there are many pebbles associated with greyish buff clay with white and brown angular flints. An old chalk pit [39925722] 500 yd S. of Bedlestead showed above the chalk 4 ft of brown clay with flints and very few Tertiary pebbles, the whole covering pockets of vermilion-coloured clay. One deep pipe in the Chalk contains vermilion-coloured fine loamy sand, angular flints and a few pebbles, and is lined with stiff reddish brown Clay-with-flints, some with manganese staining. Abundant Tertiary pebbles occur in pockets in chalky soil [397580] 350 yd N.N.W. of Bedlestead and in the Clay-with-flints about 500 yd E.S.E. of it; 500 yd S. [39925723] remanié Blackheath Beds consisting of closely-packed pebbles in a loamy sandy matrix lie in pockets in the Clay-with-flints. Fine bright reddish brown loamy sand associated with few pebbles but many angular flints occurs at the top of Botley Hill and about ½ mile N.W. of this [392559] there are pockets of pebbles and yellowish pink sandy clay. At Betsom's Hill there are many pebbles and ¼ mile N.W. Clay-with-flints was noted to a depth of 12 ft in old pits [430567]. In a chalk pit [44005672] in Middle Chalk, 900 yd S.S.W. of Gray's Farm, Clay-with-flints pipes, in places with a core of red sand and a few pebbles, have been exposed to a depth of 30 or 40 ft.

Irregular patches of bright red sand are included in clay west of Aperfield Court, and pebbles occur at Single Street. Pipes of Clay-with-flints enclosing red sand with pebbles were noted in the chalk pits 600 yd S.S.W. [44205648] and 300 yd S. [44485666] of Gray's Farm, and pebbles 450 yd N.E. [453573] of Silversted and at Horns Green [451587]. Scarlet and crimson sands are to be found in the Clay-with-flints near Cudham church and brown sand and loam with pebbles and subangular flints in small patches throughout the Clay-with-flints around Lett's Green.

On the spur [46855972] 600 yd W. of Blueberry Farm, yellow sand and blocks of sandy ironstone are associated with the Clay-with-flints. In the banks of the lane [47205855] ⅓ mile S.E. of Knockholt church, pink sand and thick red clay with large flints were noted, and just south-east of this brown loam with many small flints. Bright pink sand in the Clay-with-flints appears near the edge of the Chalk escarpment [49405895] about ¼ mile N. of Old Star House, formerly The Beacon; it probably occurs in pipes in the Chalk. On the heath [502604], ¼ mile N.W. of the Polhill Arms the soil contains many brown flints, together with some ironstone. At Pilots Wood [507607], ¼ mile N.E. of the Polhill Arms, the clay includes many pebbles and also, a little farther north-east, fragments of chert.

East of the Darent, coarse orange sand is present at the crest of the Chalk escarpment north of Otford Court, formerly Beechy Lees. A small chalk pit [55025953] about ½ mile E. of Otford Court, formerly clearly exposed several pipes of two types; H. G. Dines noted one type filled with reddish clay containing fresh flints with buff cortices, and lined with 3 to 6 in of black or dark brown clay with black-stained flints, and another filled with pink to white very fine loamy sand, with an inner lining of 6 in to 1 ft of buff loamy clay and sharp flints with buff cortices and an outer lining of black clay with black flints. Pipes of the latter type are overlain by 2 ft of normal Clay-with-flints and soil.

Vermilion-coloured sandy loam was noted 1000 yd N.W. [55786052] of Cotman's Ash and grey and yellow mottled clay, with patches of pebbles, about ¾ mile N.W. [55636071]. Yellow and red sand occurs irregularly in red clay around Drane Farm [576604].

Brown Clay-with-flints, and some Tertiary pebbles, in pipes in Middle Chalk were exposed in a small pit [59225970] by the roadside at White Hill, over ¼ mile N.W. of Newhouse Farm, and pipes with Clay-with-flints including sand and many pebbles in a road cutting [604599] at Wrotham Hill, ⅔ mile N.W. of Wrotham church. Normal Clay-with-flints is also present piping the Middle Chalk in the large pit [608598] about 700 yd N.W. of Wrotham church. Rich brown clay with large flints and patches of Tertiary pebbles occurs in pipes and in slipped masses on the scarp face in Green Hill Wood [613598], south of Wrotham Hill Park.

An analysis made by C. O. Harvey (Lab. No. 1030) of a sample of red clay that forms the matrix in normal Clay-with-flints, slightly weathered, from the road cutting on the Wrotham–London road, ⅔ mile N.W. of Wrotham church, shows the following composition, in percentages: SiO_2 65; Al_2O_3 15; total iron calculated as Fe_2O_3, 6·9; MgO 0·6; CaO 0·5; TiO_2 0·6; P_2O_5 0·1; loss on ignition 10·6. The sample contained only traces of carbonate and sulphate of calcium. S.C.A.H.

ANGULAR CHERT DRIFT

Pits on the escarpment crest, working the drift and solid together, may be up to 15 ft deep; the chief exposures are on Toy's Hill, above Hanging Bank, south-west of Apps Hollow, formerly Apps Bottom, and between River Hill and Carters Hill. The drift capping the subsidiary escarpment of the Folkestone Beds, near Ightham, is of similar character to that on the Hythe Beds but contains fragments of Ightham Stone and Oldbury Stone (see p. 62) as well as the usual chert and ironstone. The matrix is also similar, for it is derived from the loamy glauconitic sand, in which the chert layers of the Folkestone Beds occur, which lithologically resembles that associated with the Hythe Beds cherts. No pits are now active in the Ightham neighbourhood. H.G.D.

Cherty loam and red clay occur as pockets in solution hollows, 6 ft or more deep and 3 to 4 ft wide, in the Hythe Beds in the large quarries at Offham (p. 75). Brown (1941, p. 14) considers the red clayey matrix of the drift at these quarries to be analogous with that of the Clay-with-flints. Festooned and contorted beds, containing chert fragments, are frequently concentrated in the hollows but not at the bottom of the deepest pipes, while in places there are subsidiary white flints. Abundant rotten glauconite in the red clay lining the pipes suggests disturbed Sandgate Beds. S.C.A.H.

LIMPSFIELD GRAVEL

East of Limpsfield, on either side of the main road near Broomlands Farm [421533], patches of gravel with maximum thickness of 3 ft lie on an uneven slope of Folkestone Beds. Tertiary pebbles and angular and subangular flints predominate in a sandy matrix; ironstone, however, is more evident than in the Limpsfield deposit, often occurring as well rolled lumps. Pieces of polished chert are abundant and occasional quartzite pebbles. In the gravel some ½ mile S.W. of Moorhouse Bank conglomerate has been found [423527].

South of Moor House is another cap of similar gravel, but with smaller Tertiary flint pebbles than in the above, at a height of 486 ft above O.D. [430530]. Some 200 yd S. of Westwood Farm [428537], stiff grey and buff mottled clay with Tertiary pebbles and flints overlies clay without stones, seen to a depth of 5 ft. A little to the south, at a lower level, the ridge is capped by a gravel, with a small amount of clay matrix, resembling that of the higher ground and also of Limpsfield. Gossling (1941a, p. 313) described an upper gravel characterized by Tertiary pebbles, angular and subangular flints, comminuted flints, mudstones and Gault fossils in a clay matrix, separated by a brickearth from a lower gravel containing, in addition to the flints and pebbles, Lower Greensand pebbles, chert, quartz, cemented sandstone, and fragments of Wealden siltstone in a sandy matrix. The lower gravel he regarded as part of a river terrace, a view confirmed by an exposure opened in 1947.

Immediately south-west of Covers Farm [432536] is another small cap of gravel, up to 4 ft thick, composed of pebbles, flints and Lower Greensand material in a sandy matrix. The whole deposit, according to H. W. G. Williams (*in litt.*) who saw the exposure before 1892, has a general appearance of having been washed or slipped down the northerly slope.

At Farley Common [437540], over ¼ mile N.E. of Covers Farm, a small tableland of gravel, 478 ft above O.D. bears a general similarity to the Limpsfield deposit. It consists mainly of Tertiary pebbles and angular and subangular flints, but the former are less conspicuous than at Limpsfield. Conglomerate, quartzite and ironstone have been found. The upper part of the deposit is unstratified and has a scanty matrix of mottled grey and buff clay. The lower part has an abundant matrix of sand and rests on disturbed Folkestone Beds.

Small patches of similar gravel occur on the south side of the valley in Squerryes Park [443533], west of Dunsdale [452539], at Valence [459542] and west of Heverswood [467545], all above 400 ft O.D. and resting on Folkestone Beds. Farther east, near Birchfield [480545], a patch of gravel described by Topley (1875, pp. 192–3) is remarkable for being interbedded with disturbed Folkestone Beds. The gravel, about 3 ft thick, consists of some 60 or 70 per cent of Tertiary pebbles, mostly small; the rest of the deposit, comprising well worn blue flint nodules, battered and broken brown flints, chert, sandstone, rag, and ironstone with a small amount of sand matrix. Farther east [490548], capping ground over 400 ft above O.D. 350 yd E. of Sundridge Place, lies another small patch of gravel.

Three patches of gravelly material at Salters Heath [510550] lie between 445 ft and 465 ft O.D. The gravel is 4 to 5 ft thick and consists of accumulations of somewhat abraded Lower Greensand stones in a matrix of brown sandy loam.   S.B.

SCARP DRIFT AND DOWNWASH

Brown-weathered clay, about 3 ft thick, with unsorted brown and white flints, irregularly caps the Gault in a pit [433541] ¼ mile E.N.E. of Westwood Farm, near Westerham (see p. 89) and a sprinkling of drift was noted by Prestwich near Ivy House, 1 mile N. of Westerham. A borehole [47055570] 350 yd N.E. of Brasted church passed through 10 ft 3 in of drift overlying the Gault. Beneath the top soil the material consists of yellow sandy clay with streaks of blue clay and small chips of flint. A similar deposit occurs at Greatness, near Sevenoaks, and again on the north and east sides of the pit of the Sevenoaks Brick Works Co. (see p. 91). Heavy loam and brickearth up to 8 ft thick, with flints in places and re-sorted greenish loam probably from the base of the Gault, occupies a low-lying area of Gault and mantles the Folkestone Beds slopes a little southwards.

A good deal of chalky rubble fills the escarpment coombes west and north of Chevening but, as also farther west in the Westerham–Titsey area, only a thin flinty wash persists southwards on to the Gault. Flinty loam with chalk mantles the Lower Chalk north-west of Dunton Green. The left bank of the Darent near Chipstead and Riverhead is partly covered by gravelly drift with chalk, associated with much-disturbed clays and sands at the base of the Gault.

In the Chevening district, Prestwich (1891, pp. 147–51) has described drifts between Combe Bank, near Sundridge, and Dunton Green. Of three patches indicated on his sketch map, however, only the middle one, south-east of Chevening Cross, is significant as a continuous spread. His description indicates that the drift is festooned irregularly into the Gault Clay and consists largely of angular unworn flints. He described the material seen in the railway cutting "between Chevening Cross and Combe Bank Wood" as unstratified gravel, sand and clay: the lower part coarse, in a matrix of ferruginous loam and sand, mixed with some clay from the Gault; the upper part, spreading uniformly over the whole, a light brown clay (altered Gault), mixed with a few flints. In postulating a return to a glacial climate to account for the sludging effects Prestwich anticipated the modern view that some head deposits of this type may have accumulated or have been disturbed under periglacial conditions. A little gravelly drift admixed with clay occurs bordering the Alluvium ½ mile E. of the disused Chevening Station [503566].

On the west side of the Darent valley near Otford a temporary section, about 10 ft to 12 ft deep at 280 ft O.D., near Sepham Farm, was recorded by Prestwich (1891, p. 155) as follows: "Red argillaceous rubble with dispersed Chalk flints and Tertiary pebbles. Chalk-rubble of broken chalk and sharp angular flint-fragments in a chalk-paste, passing into—Solid Chalk, with layers of flint". In the chalk-rubble were found occasional Tertiary pebbles and a few fragments of Lower Greensand ironstone and chert. Absence of sorting and the presence of items probably derived from the heterogeneous materials in the Clay-with-flints, suggest the deposit to be a local head.

Chalky Scarp Drift exposed in a chalk pit east of Otford railway station was described by Kennard (1897, p. 209) and by Chandler and Leach (1909, p. 236) as a chalky rainwash mud, up to 8 ft thick but thinning off towards the scarp, with granules and pellets of chalk and a small proportion of earthy matter; other contents include pebbles, broken flint, chert and well-preserved shells of land mollusca in great abundance. Further chalky drift is to be found on the north side of the Pilgrims' Way eastwards below Otford Court and a buff stony hill-wash was noted in the lane [54645888] ½ mile S.S.E. of Otford Court, and in the hollow [55205915] about ⅔ mile E.S.E. of the same locality. A Recent drift of chalk rubble and disturbed Clay-with-flints material, probably produced by soil-creep processes, also overlies solution pipes and extends down the hill-slopes at the old chalk pit (Shore Hill Chalk Quarry [549594] see p. 115) just over ⅓ mile E.S.E. of Otford Court (Leach 1921, pp. 38–9).

Another thin spread of drift, about ½ mile E.S.E. of Otford Station, and about 300 to 330 ft above O.D., is, according to Prestwich (1891), an angular and sub-angular white flint gravel with a few brown-stained worn flints and Tertiary pebbles and a very few fragments of chert and rag. The spread seems to be irregular and to lie in pockets in Chalk rather than to be a uniform bed, as do two patches on the Gault mapped by Prestwich near Child's Bridge, south-west of Kemsing. The first of these, north of Child's Bridge [545580], is described as angular flint-debris, probably of the date of Prestwich's middle series, his Chevening and Dunton Green Gravel; the second, classed with his Low Level Valley Gravels, lying along the railway to the south, consists of ochreous and roughly stratified gravel, 4 to 5 ft thick in places, with a preponderance of Lower Greensand debris and a few flints and pebbles.

Over 3 ft of buff chalky wash were seen on the north side of the Pilgrims' Way [562592] ⅓ mile N.W. of Crowdleham House and some of this wash persists southwards on to the Gault near Heaverham. A certain amount of clayey downwash with small white flints occurs farther south on the Gault, towards the railway, but only two patches were uniform and thick enough to warrant mapping. These lie on the watershed south-east of Heaverham, the more southerly patch [57705725] showing over 6 ft of drift where cut by the railway.

Near Fuller Street [561568], Stonepitts [569570], Broomsleigh [575567] and north-ward from these localities a hill wash of sandy and loamy rubble, with angular and subangular lumps of chert and some sandstone, mantles the Folkestone Beds slopes and extends on to the Gault. In valleys it is up to at least 8 ft thick. Prestwich, though tentatively classing this Drift with river drift, nevertheless states (1889, p. 276 footnote) "some of this may, like the drift around Oldbury, be local debris or trail from the higher ground above".

Prestwich described striking instances of the erosional force of heavy rainfall noted by Benjamin Harrison in 1888 when coarse flint rubble to a depth of 2 to 3 ft was thrown across the Pilgrims' Way near St. Clere by torrential water from a coombe in the Chalk scarp. Harrison had recorded this remarkable deposit in notes in his diary (see Harrison 1928, p. 135).

Chalk and flinty wash is found in variable amounts on the Middle and Lower Chalk eastwards of St. Clere towards Wrotham. It is mainly a gravel of white angular and subangular flints, large unworn flints, a few brown-stained subangular flints, Tertiary pebbles and a very few fragments of Lower Greensand chert, rag and iron-stone. The most significant spread irregularly caps the Lower Chalk spur [582589], at about 430 ft O.D., between Yaldham and St. Clere. The irregular nature of the drifts occurring locally on the Gault outcrop farther south-west is well demonstrated by the records of a series of trial holes put down in the area north-west of Ightham Court, also mentioned below (pp. 130-1). Seven of the records show Gault at the surface and eight, mostly in valleys and hollows, prove thicknesses of between 2 and 6 ft of gravelly drift.

On the Folkestone Beds north of Ightham, loamy drift with flints and chert forms a hill-wash similar to that farther west. Several drift patches merit separate consideration, since hereabouts and to the south of Ightham much of Benjamin Harrison's work on the drifts and their implements was carried out. Bennett (1908b, pl. II) and Prestwich (1889, pl. IX) show various drift patches here, and the latter's map summarizes the implement-yielding localities and certain names of drift patches important in con-nection with Benjamin Harrison's work on the flint implements of the whole district.

A drift patch on the northern slopes of Oldbury Hill [585570], a little above the 400 ft contour, consists mainly of sand and loam with chert, ironstone and locally-derived hard quartzite. Prestwich realised that local debris from higher ground had contributed to this drift, which he classified with his High-level or Limpsfield Stage. Two small patches of more flinty loam are shown on the lower slopes of the hill [590569] about ⅓ mile E. (Fane Hill and Bayshaw localities of Harrison), but north of Ightham church (Coney Field locality of Harrison) and around Ightham Court (Court Lodge locality of Harrison) only pockets of flinty loam represent these Lower-level Old Shode gravels, which Prestwich recorded as apparently 2 to 4 ft thick (4 ft at Bayshaw). A more important gravel at High Fold [600568], about ⅓ mile E.S.E. of Ightham church and over 300 ft above O.D. is, according to Prestwich, not less than 8 ft thick, its stony content consisting of 50 per cent of subangular white flints, together with a few brown ones, 45 per cent of Lower Greensand debris, and 5 per cent of Tertiary pebbles. Prestwich suggested that torrential streams in late Glacial time may have accounted for much of the debris from the scarp in the 'Shode drift-gravels', the rest being derived from the remnants of an old unstratified gravel on the Gault in the Vale of Holmsdale. The brickearth to which he refers "on the line of the railway just below Fane Hill at 300 ft" appears to be that now included as loamy Alluvium.

Three other gravelly spreads lie north-east of Ightham. The first is a sandy wash, with many flints, over 4 ft thick and occupying the hollow [597573] ¼ mile S.E. of

Ightham Court. The second caps the hill $\frac{1}{3}$ mile E. and was well seen in workings [602575] $\frac{1}{2}$ mile N.E. of Ightham church (see p. 85). The average thickness is 6 ft but the junction with the Folkestone Beds is very irregular and pockets of angular gravelly drift extend to a depth of 12 ft. On the north side of the pit the flints were seen to be characteristically irregular, large, white and somewhat battered; a few Tertiary pebbles occur and the matrix consists of rearranged Gault material, with fragmentary fossils. The drift showed practically no stratification. On the south side of the pit, and in a small cutting parallel with the railway line, subangular white flints were seen to be partly concentrated into thin beds and lenses, festooned into a deep broad pocket. The matrix here is a brown sandy loam and fragments of chert and rag are more abundant. The material of the third patch of drift [605576], about $\frac{1}{4}$ mile N.E., resembles that on the north of the pit.

Along the Pilgrims' Way between Newhouse Farm and Wrotham chalky drift is considerable but discontinuous. A section seen by Bennett showed from 3 to 6 ft of fine chalky drift with a layer of unworn flints at the bottom and a few land shells, resting on Lower Chalk. Southwards of Wrotham a little thin brown clayey wash with flints occurs occasionally on the Gault, but mainly there are only scattered unworn white flints, which cannot be regarded as amounting to a spread of flint referable to "an old flood plain of the Shode" as decribed by Bennett (in Bennett and Harrison 1906, p. 462). Eastward of Wrotham drift occurs in the deeper coombes, as along the Pilgrims' Way [623596], $\frac{1}{2}$ to $\frac{3}{4}$ mile E.N.E. of the church, and near the junction of the Lower Chalk with the Gault. Marly wash, locally with land shells or white flints, lies about the spring heads at Little Wrotham [62565926] and Wrotham Water [629597], between Wrotham and Trottiscliffe. This chalky drift extends in places on to the Gault, as at Moat Farm [62505893], where it is 2 ft 6 in thick, and again [64776004] about $\frac{1}{3}$ mile E. of the inn at Trottiscliffe. About $\frac{1}{4}$ mile S. of Trottiscliffe the soil is very flinty and in a borehole at the Mid-Kent Waterworks [64025956] 2 ft 6 in of soil on 2 ft 6 in of gravel overlie the Gault. For the most part, however, the Gault seems bare of drift hereabouts. East of Trottiscliffe, near to Ryarsh, though there is very little wash on the Gault, much brown loam with angular flints and lumps of ironstone covers the lower slopes of the Chalk escarpment and obscures the boundary between Lower Chalk and Gault near Trottiscliffe church.

Around Borough Green and to the east, a continuous spread of Drift on low ground is essentially a brickearth, but gravelly seams are locally present (Plate VIIIB). In the sandpits north of Borough Green railway station (see p. 86) the drift was well developed in the three pits nearest the road to Wrotham, but in the westerly pit [607578], $\frac{1}{4}$ mile W. of Longpond, the amount of superficial material was negligible. The most southerly pit [60855760], now worked out, formerly showed 6 ft of brown and mottled sandy loam and brickearth with scattered flints, etc., concentrated into a bed 6 in to 1 ft thick near the base. The contents include angular and subangular flints, ironstone, occasional quartzite from the Folkestone Beds and white sandstone derived from seams in the latter occurring in the pit. There is no regular bedding but the drift showed complicated festooning, especially in deep pockets extending downwards about 10 ft. In the next pit [60955780] to the north-east the drift section was as follows: made ground up to 3 ft, light sandy loam with a few scattered flints, etc., 3 to 4 ft, and coarse angular to subangular gravel in a sandy matrix, 2 ft; the gravel contains flints large and small, mainly white and a few Tertiary pebbles and small lumps of Lower Greensand ironstone; deep pockets increase the thickness of the gravel.

In the northerly pit [610579], near the Gault outcrop, glauconitic clay with phosphatic nodules is admixed with a heavy brickearth here overlying the flinty layer; in the Gault itself on the north side of this pit there are pockets of coarse angular to subangular gravel. Brown (1928, p. 194) and Holmes (1937, p. 351) noted variations in the drift at these pits, and it is recorded by Bennett that in 1906 a section, now obscured, in old Gault clay-workings just north of the present sandpits, showed brown loam passing into pelletty chalk drift from the escarpment. The loam was considered to be the result of decalcification of the chalky drift; other downwashes from the south, however, predominated in its accumulation.

Other sections in the same loamy drift have been described at the brick and tile works [621577] about ¾ mile E.N.E. of Borough Green railway station, where the drift is more uniformly of brickearth type. In the low-lying ground here a number of shallow excavations where the brickearth had been dug exposed brown and slightly mottled loamy brickearth with some small white flints and pieces of ironstone, the latter mainly on the south side. Passing downwards the flints become scarcer and the drift is a brown sandy clay without debris. Below this lies a thin bed of flinty sand overlying the Folkestone Beds, the total amount of drift on the solid not being more than 4 or 5 ft; another section showed 1 ft of sandy wash with ironstone, a few small flints and tiny pieces of chalk, overlying 4 ft of brown loam with ironstone and a few flints, grading into a heavy dark brown loamy brickearth, seen for 3 or 4 ft. This drift tails off up the side of the hill to the south and merges into sandy soil. An analysis made by C. O. Harvey (Lab. No. 1031) of a sample of brickearth taken from 6 ft below the surface, near the valley bottom, is as follows, in percentages: SiO_2 82; Al_2O_3 7; total iron calculated as Fe_2O_3 3·5; MgO 0·7; CaO 0·3; TiO_2 0·6; P_2O_5 0·1; loss on ignition 4·2. The sample contained only traces of carbonate and sulphate of calcium. On higher ground just to the north variable amounts of large white angular battered flints occur sporadically mixed with clay which is largely disturbed Gault. Moreover, passing just southwards on to the Folkestone Beds the drift changes to a sandy wash (2 to 3 ft) with white flints and ironstone. Prestwich grouped this drift with his Higher Unclassed Gravels.

Other nearby drifts are discontinuous and local. Above Dark Hill Sand Pit [60455715], ⅓ mile W.S.W. of Borough Green Station (see p. 86), 3 ft of sandy wash with a few flints but much ironstone were seen to tail off up the hill to the south. In the sandpit [612570] near the Black Horse, about ¼ mile S.E. of the station (see p. 86), Bennett recorded pockets of Lower-level Brickearth with pelletty chalk, unworn flints and Tertiary flint pebbles. The overburden in the sandpit [622572] 200 yd N. of Platt church (see p. 87), although only 1 to 2 ft thick, also contained small pieces of chalk, with white flints and ironstone. Lumps of ironstone and a few small white flints were noted in sandy wash above a sandpit [62375760] about ¼ mile N.N.E. of the last. Eastward of Platt sandy soil contains ironstone and chert, with a few small white flints—the latter rather more abundant (High Level Gravels of Bennett) at Gallows Hill [630578] and around Highlands [638574]—also brown flints, Tertiary pebbles, and, south-east of Highlands, much ochreous flint and a few blocks of Oldbury Stone.

In the valley north of Wrotham Heath is a loamy brickearth with sandy wash overlying it in some places. It was seen in the north-west part of the sandpit [63035814] 250 yd S.E. of Nepicar House (see p. 87), being there exposed to a depth of 6 ft and containing locally-derived lumps of sandstone from the Folkestone Beds.

On the higher ground in the Wrotham Heath area white flints were seen in places in thin sandy wash and occasional loamy pockets in the Folkestone Beds. Similar sparse distribution occurs in the area of Addington Park and north-eastwards of Ford Place and of Westfield Farm.

In the Ryarsh, Addington and West Malling neighbourhoods the drift is divisible into a gravelly type on high ground, often dissected into strikingly terrace-like hill-cappings, and a brickearth type occupying valleys and mantling hill-slopes. The former is the older and the brickearth near Ryarsh is in some places connected as a continuous spread with the gravelly drift, from which it incorporates a very flinty downwash.

The most important local occurrence of gravelly drift is that south-west of Ryarsh, where coarse angular and subangular gravel, essentially of flint in a sandy matrix but with occasional ironstone, was seen along the road from Ryarsh to Trottiscliffe, and, over 6 ft thick, in old pits in Dunstan's Rough [667595], about ¼ mile S.W. of Ryarsh. The gravel is rough, unworn, unsorted and unbedded. This drift, recorded by Prestwich, was also formerly well seen in the large sandpit [66905955] just south of Ryarsh (see p. 88). In 1947 only variable irregular loamy brickearth with sub-sidiary unworn flints was exposed (north-west part of the pit) but previously brown sandy loam, crowded with flints in beds or festooned in irregular masses, was to be seen overlying the Folkestone Beds and varying in amount from 1 to 9 ft.

The drift is also very gravelly farther south-east, towards the church, and at about the same level is a capping of drift at East Street, over ⅓ mile E. by N. of Addington church. A small roadside section [65975905] there showed 5 ft of sandy drift with flints, ironstone and Kentish rag, resting on a 6-in basal bed of small subangular flints and ironstone pebbles overlying fine sand of the Folkestone Beds. Harrison noted a large spread of pitted and ochreous flint in this region, but most of this drift is a superficial wash. It is also thin and irregular about Addington church. Westward of Woodgate a sandy wash includes tiny pieces of chert and some flints. Traces of gravel in pockets in sand of the Folkestone Beds occur as a hill capping in Leybourne Wood [684585], ½ mile S.S.E. of the Mental Hospital, formerly Leybourne Grange. Topley (1875, p. 174) regarded this as an outlying patch of Medway gravel. The level of the highest point of the hill, 153 ft above O.D., is approximately that of neighbouring spreads.

On the south side of the stream at Addington, gravelly drift with coarse sand and ironstone forms a thin capping to the flat-topped hill [671588] bordering the Maidstone road ¼ mile S.S.W. of St. Martin's Church. A certain amount of flinty wash persists north-eastwards from this patch. Very flinty loam, the flints mainly white, also caps the hill [666583], at a higher level, ½ mile N.N.W. of Fartherwell. Pockets of small flints occur just north of Offham church [66055812] and on the hill [655584] ½ mile N.W. of this, scattered flints and pieces of ironstone, chert and ragstone are strewn over the loamy and sandy soil of the Folkestone Beds; a little brown loam with small white flints occurs near West Malling, 150 yd N.E. of the railway cutting. The spread of sandy drift, with subangular flints, chert and Tertiary pebbles, which Prestwich noted at the level of 230 ft is discontinuous, and only a thin uneven loam extends to West Malling.

The brickearth type of drift in the Ryarsh, Addington and West Malling district is widespread, but sections are few. In its extensive development near Birling and Ryarsh and southward near the Mental Hospital the drift is variably flinty and loamy but flints are particularly abundant on the west, near the gravel patches described above. Where it tails off to the north it becomes coarse and sandy and at Birling it passes imperceptibly into a sandy wash with flints and chalk fragments. A similar chalky loam was seen a little below the Gault outcrop ¼ mile N.N.E. of Ryarsh, where an exposure [67066021] showed 3 ft of earthy loam with flints and chalk overlying a 1-ft bed crowded with chalk pellets and containing some flints. Temporary exposures [678592] near the Mental Hospital showed up to 6 ft of loamy brickearth with flints on the Folkestone Beds. To the south this drift contains ironstone and is again very sandy. In the quarry [679585] ⅓ mile N.N.W. of West Malling (see p. 79) sandy wash contains angular and subangular white flints and ironstone fragments with a facetted appearance, perhaps wind-worn; small flints are concentrated near the base. It is thickest on the valley slopes, and varies from 1 to 5 ft towards a tributary valley on the east. A small tongue of loamy drift extends northwards a little up the valley [662590] east of East Street, ¼ mile E. of Addington.

South of the stream near Addington similar drift is banked against the hill slopes near the railway and the Maidstone road. It was seen to be at least 6 ft thick in some places.

West of West Malling the Hythe Beds are overlain by a wide spread of rich brown loam of a brickearth type, of variable thickness and lying largely in pockets; it appears to be a residual product of denudation of the Hythe Beds. Also near West Malling and again north of Mereworth, north of Plaxtol and near Claygate Cross, loam mantles the Hythe Beds slopes below the Angular Chert Drift. In the old quarry [67805835] ½ mile N. of West Malling church an overgrown section showed the drift to be 3 or 4 ft thick and to include with it some flinty wash.   S.C.A.H.

HEAD DEPOSITS ON WEALD CLAY

Group one:

South of Limpsfield, at the base of the Lower Greensand escarpment, one of the highest occurrences of pebbly drift on the Weald Clay, lying some 350 ft above Ordnance Datum, is composed of well-rolled blue flint pebbles, battered brown flints and subangular fragments of Lower Greensand chert, sandstone and ironstone in a sandy clay matrix. Similar material was found at Hurst Green, 300 ft above O.D. on the eastern side of the stream which flows south of Coltsford Mill, and ½ mile S.E. of Hurst Green, at the same level within the drainage area of the Kent Brook, as well as on the watershed of these streams, between Foyleriding and Highridge farms, also 300 ft above O.D.

To the south-west and south-east of Highridge Farm similar drift occurs at lower levels in the valleys where it is distributed like remnants of a terrace. The deposit is thin but in old pits [411463] ½ mile E.N.E. of Chellows Park subangular brown flints, blue flint pebbles, rounded buff sandstone, chert and ironstone and lumps of conglomerate were seen to be embedded in a sandy clay matrix. Drift of this type again occurs at Haxted, Skeynes, ¼ mile S.W. of Medhurst Row, Whistlers and How Green. These spreads are about 200 ft above O.D. Flints were abundant in the soil on the southern margin of the Camp Hill outlier and have been noted at the northern end. A section by Bird (see below) in the centre of this outlier has shown that the Head is here free from flints. It is possible that the first two groups of Head deposits are in contact at this locality. Later downwashing of the sand and gravel has obscured the southern boundaries of the Camp Hill outlier.

Group two:

The most westerly deposit is thin, and lies in pockets. It contains Lower Greensand fragments of subangular chert and glauconitic sandstone only, and caps a small hill [431487] some 2 miles N.N.W. of Edenbridge.

On the Weald Clay below Toy's Hill and Ide Hill the drift, lying in numerous patches, is composed of angular and subangular chert and sandstone, together with fine-grained sand, of Hythe Beds origin. This now caps the low ridges between the streams which flow from the Lower Greensand to the River Eden; the patches are evidently the remains of a plateau-like spread which extended from the base of the escarpment towards the river.

Patches of drift cap low ridges south-west of Sevenoaks Weald and extend from Bore Place [507488] south along the western slope of a minor valley almost as far as the Eden just north of Chiddingstone. At its southernmost end the spread is 150 ft above Ordnance Datum, 50 ft lower than the flint drift on How Green.

An extensive deposit of the chert drift occurs at Chiddingstone Causeway. At its highest point the cap is 194 ft above Ordnance Datum and inclines east of south.

A section at Camp Hill [523468] was described by Bird (1963, p. 449) as showing 3 to 5 ft of cherts in a matrix of silty clay overlying 10 ft of contorted, weathered clay and then unweathered clay and bedded siltstone *in situ*. South of River Hill there is another extensive spread consisting of a rubble of chert and sandstone with a subordinate amount of sand, again capping low ridges. Farther east, at Home Farm [605525], between Shipbourne and Dunk's Green, the drift is composed of angular and subangular rock from the Hythe Beds, but farther south, in three spreads between Hoad Common and Starvecrow, pockets of subangular flints and pebbles were noted, although chert and sandstone predominate. Topley (1875, p. 185) mentions the presence of Wealden pebbles as well as of flints and Tertiary pebbles around Starvecrow. North of Trench Wood [588493] and at Little Trench [595485] the drift consists of Lower Greensand material.

In many places the Weald Clay surface slopes steeply to the west and gently to the east and it is on the gentle slopes and in the bottom of the valleys that traces of chert and sand are to be found. On uncultivated land or poorly cultivated pasture on these drifts, the vegetation is normally in marked contrast to that on the surrounding clay, the sand favouring in particular a vigorous growth of bracken.

Foster and Topley (1865, p. 449) described a gravel spread at Dunk's Green, as consisting of river gravel with Wealden pebbles on the south and large angular flints and Tertiary pebbles with a piece of chalk and no Wealden pebbles on the east. This comprises two portions, separated by bare clay. The western part is in fact that described above at Home Farm, in which chert predominates.

Group three:

The eastern part, lying about 60 ft above the Shode or Bourne and grouped in the third division, contains subangular cherts, battered, pitted brown flints usually retaining some white cortex, and some well-rolled blue flint pebbles. Mid-way between Claygate and Stallions Green another patch [615515] with abundant cherts, less conspicuous battered brown flints and a few Tertiary pebbles with some sand rests at about 60 ft above the stream. Further spreads, to the south-east of this last, occur at lower levels in the valley at about 20 ft above the stream; they contain seams of sand exhibiting crude bedding, and have been regarded as river terraces.   S.B., C.R.B.

Head Deposits on Hastings Beds (Hillwash Head)

Four feet of yellowish brown sandy loam, overlying silts of the Ashdown Beds were noted in the road bank by Elliott's Farm [52994357].

Two small valleys south of Haysden [569456] are infilled with Head. Northwards the Head extends on to the floodplain of the Medway and merges with the Brickearth. A trial borehole [57424508] in connection with the proposed Tonbridge By-pass demonstrated that the Head is present to a depth of at least 22 ft where it rests on 2 ft of gravel. Lower deposits of firm blue, grey and brown silty clay to a depth of 41 ft are tentatively classified as river deposits but may represent an earlier period of solifluction. Gravel is present from 41 to 47 ft and the lowest 2½ ft down to 60 ft O.D. consist of stiff brown and blue sandy clay resting on Ashdown Beds. Farther east, towards the margin of the Head, another borehole [57764512] proved 24 ft of soft brown sandy clay with sandstone fragments at 102 ft O.D. resting on Ashdown Beds. At 90 yd E. [57804512] 9½ ft of Head rest on the Ashdown Beds at 125 ft O.D.

Head in the valleys in the eastern part of the map area does not appear to be so thick as in those immediately adjacent to the Medway. Locally shales of the Wadhurst Clay are exposed in the stream bed, although the sides of the valley are completely masked by Head, which can be proved by augering to be at least 4 ft thick. Five feet of brown sandy loam were noted in the bank of the path [63154311] leading to Kent College, formerly Hawkwell Place.   C.R.B.

Terraced River Gravels

Eden and Medway Valleys. Between 4 and 8 ft of First Terrace gravel exposed in temporary trenches in Edenbridge showed patches of loam, up to 3 ft thick and containing small flints and chert fragments and small Tertiary pebbles, overlying a coarse gravel of flints, pebbles and chert in a sandy matrix.

Patches of coarse sandy flint, pebble and chert gravel of the Second Terrace at about 170 ft above O.D., were exposed in the railway cuttings north-east and east of

Edenbridge. Included in similar gravel about 1 mile E. of Edenbridge are well-rolled brown and blue flints, Tertiary pebbles, subangular and well-rounded cherts and cherty sandstone, and ironstone.

South of the Eden between Lingfield and Edenbridge loam and gravel of the First Terrace overlie the Upper Tunbridge Wells Sand. Ditch sections [401447, 40264470 and 406448] to the north-east of Park Farm showed 2 to 3 ft of sandy loam overlying gravel composed of small pebbles of sandstone and ironstone. The southern feather-edge of the terrace is much broken by frost-heaving and the gravels are thin and festooned into the underlying Tunbridge Wells Sand.

Murchison (1851, p. 381) described gravel at Hever Lodge Farm [482462], 1 mile N.N.E. of Hever church, as consisting "of small flints of yellow, red and white colours, the greater number angular, and fragments of clinkers and chert of the Greensand, with some rounded black pebbles of flint". He also noted that the stones are mixed with sand and loam which is locally ferruginous enough to cement the gravel. The pits [484459], which lie between Hever Lodge and the river, are now mainly over-grown, but the thickness of the deposit, classed with the Second Terrace, is estimated at 7 ft.

Immediately east of Chested [50754620], a roadside pit on the Second Terrace penetrated 2 ft of brickearth into 8 ft of small gravel resting on coarse gravel seen to a depth of 2 ft. Some 250 yd N. of Penshurst church, in a small pit [52774425] also on the Second Terrace 30 ft above Alluvium level, 2½ ft of brickearth were seen to overlie 2½ ft of coarse gravel, on fine gravel mixed with quartz-glauconite sand, seen to 3 ft.

In a roadside pit [54904517] at Ensfield, 3 ft of brickearth were seen to rest on 6 ft of sandy gravel, consisting of small Tertiary pebbles and fragments of flint, chert, ironstone and Wealden sandstone. The base of the gravel was not exposed but its surface is about 30 ft above the Alluvium and is accordingly correlated with the Second Terrace. Among gravel occurrences described by Topley (1875, p. 185) is a 5 ft section at Ensfield.

At The Stair [60904765], a section on the Second Terrace at about 30 ft above Alluvium level showed over 7 ft of brown and olive green sand with seams of flint and gravel of ironstone pebbles wedging into current-bedded gravel at the base. The gravel bed passed under Brickearth to the north-east and is at 16 ft depth near the twelfth milestone on the Tonbridge–Hadlow road.

Almost continuous with this spread, similar gravel forms a plateau at Great Fishall Hall [618479], where 8 ft of sand and gravel were exposed in an old pit. Near Faulkners and ½ mile N.E. of Hadlow church, other patches of Second Terrace consist of sand interbedded with gravel at a slightly higher level.

A different type of gravel occurs in adjoining patches of Second Terrace drift. For example, along the Hadlow road about 300 yd N.E. of Faulkners, road widening [62504935] showed loam with large subangular flints and cherts. Farther north, just west of Bourne Grange, a similar spread of flint and chert gravel 2 ft thick and free from sand overlies Weald Clay.

On the eastern side of the tributary stream to the Medway, a section in an old pit [630503] showed 4 ft of crudely bedded coarse gravel of the Second Terrace consisting of subangular flints and cherts with some loam, locally with ferruginous cement. C. E. Hawkins (in manuscript) here recorded 5 ft of coarse chert and flint gravel, resting on 4 ft of fine gravel consisting of chert, flint and Wealden debris, lying on coarse sand. Roadside pits [628511] about 1 mile N.W. of Hadlow exposed 6 ft of roughly bedded gravel of the same terrace lying 30 ft above the river. The drift consists of angular and subangular brown flints showing white spotting, angular and sub-angular brown cherts, glauconitic and cherty sandstone, ironstone and a few Tertiary pebbles, usually broken; it contains sandy seams and has a clayey sand matrix. A patch of similar gravel occurs at the same level by Stallions Green, on the western side of the valley.

In the gravels lying at greater heights sections were only seen in the spread of Fourth Terrace gravel 1 mile N.N.E. of Hadlow. On the northern face of the westerly of two pits [641511] the section showed: grey clayey soil with pebbles, 2 ft; iron pan, 2 in; yellow sand, locally clayey, with large subangular ill-sorted flints, 2 ft 4 in; iron pan, 2 in; wisps of sand, 4 in; gravel of less angular flints and chert, small dark ferruginous pebbles and pieces of Ightham Stone, 2 ft 6 in; 'pepper and salt' sand, 6 in; grey sand and gravel with laminae of sand, 2 ft 6 in; bedded sand, 1 ft; gravel, seen to 2 ft 6 in.

The lower 6½ ft, in which there are fragments of Wealden sandstone, are more regularly stratified than the beds above. In the pit [642511] to the east, 11½ ft of sand and gravel were seen to overlie Weald Clay. Topley (1875, p. 184) noted hereabouts a section in 10 ft of gravel interstratified with false-bedded sand.   S.B., S.C.A.H., C.R.B.

Darent Valley. The spread of Fourth Terrace gravel east of Broughton House, Dunton Green, at about 150 ft above river level, was described by Topley (1875, p. 189) as distinctly and regularly stratified. Prestwich (1891, p. 142) correlated this occurrence with gravel patches at Sundridge, Farley Hill (near Westerham) and Limpsfield Common which are now classed as Head. The two remnants of Fourth Terrace gravel at about 350 ft O.D. north-west of Seal consist largely of chert and ironstone of Lower Greensand origin together with some ochreous flints. An exposure in similar ochreous gravel of this terrace just south-east of Riverhead church showed a high proportion of rounded chert fragments.

Third Terrace gravel worked ½ mile W.S.W. of Brasted is loamy and unbedded on the northern slope of the hill. A small pit [51835830] 400 yd N. of Rye Cottage, formerly Rye House, near Dunton Green, showed 8 ft of partly ochreous flinty gravel, with Tertiary pebbles, in yellow sand. The large pit [546569] west of Seal exposed about 18 ft of gravel which includes coarse shingle as well as lenses and layers of false-bedded sand; the stony material is largely Lower Greensand debris, generally well abraded, with a small proportion of flint, only slightly water-worn.

A gravelly deposit in the dry valley ½ mile S.S.W. of Brasted church probably belongs to the Second Terrace stage. It was exposed at the eastern end of the large sand-pit [46605476] beneath 10 ft of loamy sand and brickearth with stones, a local Head derived largely from Sandgate Beds. In the 10 ft of ferruginous gravel seen were ochreous flints, some up to 1 ft across, Tertiary pebbles, blocks of puddingstone and sarsen-like sandstone (probably derived from the Folkestone Beds), and lumps of ironstone, chert and ragstone. Workings in Second Terrace gravel eastward of Sundridge proved thicknesses between 3 and 12 ft. Well-bedded sandy gravel worked on the north side of the river [49405613] ¼ mile W. of Chipstead underlay 5 ft of brickearth, formerly extensively dug. Topley (1875, p. 191) reported gravel resting here on rearranged Folkestone Beds at 6 or 8 ft depth. The gravel ½ mile N.E. of Riverhead, proved in a borehole to a total thickness of 17 ft, was at one time best seen at the large wet working [521567] (Plate XB) in Folkestone Beds, where its base slopes gently northwards, towards the river, and its thickness diminishes in the same direction from 12 to 6 ft. In the compact top part of the gravel are irregular masses of clay, up to a foot or more across, but the lower part is interbedded with sand layers. The stones are mainly well-worn fragments of chert, sandstone, ragstone and ironstone; flints are rare, small and but slightly abraded and Tertiary pebbles very rare. An occurrence of gravel with Lower Greensand debris recorded by Prestwich (1891, p. 151) at the former Otford Brickpit [531578], may be referable to Second Terrace river gravel, but other occurrences he noted in the vale farther east (1889, p. 274; 1891, pl. VII) are probably thin washes of Head origin.

Dredgings for ballast at Alluvium level south of Dunton Green revealed bedded chalky sand and gravel of the First Terrace, much disturbed and overlain irregularly by clayey head.   H.G.D., S.C.A.H., C.R.B.

BURIED CHANNEL DEPOSITS

In the neighbourhood of Tonbridge up to about 16 ft of ballast or gravelly sand are present in some places below alluvial clay and silt.

Trial boreholes for the Tonbridge By-pass across the Medway valley west of the town have shown that Alluvium and gravel vary from $17\frac{1}{2}$ to 23 ft but average about 20 ft in combined thickness. These rest on Wadhurst Clay at an average height of 58 ft O.D. across most of the valley. It is only on the northern border of the flood plain [56174611], where the river abuts the Ardingly Sandstone, that a greater thickness of river deposits has been proved. Here 15 ft of alluvial clay rest on 18 ft of sand and gravel. Wadhurst Clay floors the base of the channel at 47 ft O.D. South of the present flood plain, near Haysden, up to 50 ft of silty clay with intercalated gravel and a gravelly base have been proved overlying the Ashdown Beds [57424510]. The upper part of this channel is infilled with Head deposits but the lower with gravelly material, down to a depth of 60 ft O.D. This compares with the figures for the depth of the river deposits across the flood plain. The inlier of Wadhurst Clay [575457] $\frac{1}{4}$ mile W. of Brook Street could be interpreted as having formed in a meander loop of the Medway, at this time flowing to the south of the inlier. The meander was subsequently severed in the classic 'ox-bow' fashion and the Medway began to flow to the north. Solifluction with the formation of Head deposits completed the infilling of the abandoned ox-bow lake.

About $\frac{1}{2}$ mile E. of the town trial boreholes [c. 600463] proved up to $14\frac{1}{2}$ ft of gravel to a depth of 45 ft O.D. resting on Tunbridge Wells Sand. A 12-ft bed of gravel exploited by dredging in the alluvial belt of the Medway near East Peckham contained only about 10 per cent of flint (including pebbles of Tertiary origin), the remainder consisting of chert, ragstone and sandstone from the Lower Greensand and a very small amount of the nearby Wealden sandstone and shale. Most of these materials were in fact derived from the north via earlier, terraced, gravels and stages of Head deposits. Some 10 per cent of the Medway gravel is greater than 2 inches in size, 50 per cent ranges from $\frac{1}{2}$ in to 2 in and much of the remainder is still a coarse grit.

Gravels beneath the Alluvium of the Darent proved to be similar to the Medway material, but up to 60 per cent is made up of flint. Rather more than $\frac{1}{2}$ mile N.E. of Sundridge [493558], 16 ft of gravel excavated beneath 4 ft of Alluvium in 1932 contained many large boulders (up to 18 in across) of sarsen-like sandstone (?from the Folkestone Beds) and of Tertiary puddingstone, as well as large ochreous flints, blocks of chert and ironstone. One of a series of boreholes drilled hereabouts subsequently proved the base of a buried channel of sand, loam and flint, with beds of ballast, at $37\frac{1}{2}$ ft; another proved 28 ft of river gravel, coarse in the upper part and with stony debris from the Lower Greensand throughout. The minimum thickness of gravels was $5\frac{1}{2}$ ft, below 3 ft of Alluvium. Towards Riverhead, dredgings have continued in gravels and the underlying sands of the Folkestone Beds, between which the line of demarcation is frequently vague, much re-worked Folkestone sand being incorporated in the drift. Some 14 ft of flint and chert gravel, with a little ironstone, were proved at the southern edge of the Alluvium. Further dredging has been carried out west of the main road from Dunton Green to Sevenoaks.

The wide tract east of Dunton Green has been bored from time to time for exploitation of sub-alluvial gravel. Thicknesses of ballast, itself variably sandy, range from 2 to $22\frac{1}{2}$ ft, resting on Gault or, on the south, Folkestone Beds, and covered by up to 10 ft of overburden. The deepest gravel is confined within narrow channels.

S.C.A.H., C.R.B.

BRICKEARTH

Small spreads of Brickearth that border the Alluvium west of Tonbridge form low flat terrace-like features and consist of a uniform mixture of sand and clay with occasional scattered small Tertiary pebbles and fragments of flint and chert. The spread west of Tonbridge was penetrated in former workings for railway ballast 1 mile W. of the town.

The first of two trial boreholes west of Haysden [56384567] showed 9 ft of brown sandy clay with a little gravel near the base resting on 12½ ft of gravel, and the second, 160 yd S.S.E. [56474553], 5½ ft of sandy clay with a little gravel towards the base overlying 13½ ft of gravel.

East of Tonbridge, patches of Brickearth south of the Medway mostly rise from the Alluvium to a height of 10 or 15 ft; but the spread at Five Oak Green reaches 50 ft above it at the southern end. The drift is a sandy or clayey loam, locally with ferruginous gravel seams.

A temporary section [64534532] at Five Oak Green exposed 4 ft of brown loamy silt, overlying 1 ft of sandstone gravel. A pink silty clay of the Tunbridge Wells Sand floored the section. Between this section and Paddock Wood an extensive spread of Brickearth masks the solid, although locally [646456 and 660457] small inliers of Tunbridge Wells Sand crop out but do not form a topographic feature.

North-east of Tonbridge, near the 12th milestone on the Hadlow road, thicknesses of 10 and 15 ft of Brickearth were seen over gravel in temporary sections [61084805 and 61134812]. Foster and Topley (1865, pp. 447-8) described disturbed Brickearth and gravel in a pit some ¾ mile N.E.; the contortions they figure were probably effected by frost-heaving at a period later than deposition. Sandy loam near the Hermitage, south-west of Hadlow, was seen to 10 ft; hereabouts the Brickearth surface is between 40 and 54 ft above Alluvium level. From Hadlow to East Peckham an extensive spread of Brickearth slopes evenly downwards from 50 ft above river level. At Style-place, formerly Style Place Brewery [646490] a well proved gravel and sand to over 10 ft and elsewhere it is also locally gravelly and contains flints, cherts, Wealden sandstones and occasional Tertiary pebbles. Trial boreholes at East Peckham and Hale Street proved up to 13 ft of stony loam, silty and sandy clay, and beds of gravel, overlying Weald Clay.   S.B., S.C.A.H., C.R.B.

In the Darent valley about 5 ft of pinkish buff loam partly overlie Second Terrace gravel on the north side of the river west of Chipstead (see p. 137); the loam was dug for brickmaking. Deposits at a number of localities in the neighbourhood of Seal, Riverhead, Sundridge, Westerham and at Limpsfield Common, recorded as brick-earths by Topley (1875, pp. 192-4) and Prestwich (1891, pp. 145-7), are essentially either disturbed and re-worked Sandgate Beds, or wash therefrom, or sometimes rough stony loam Head and Angular Chert Drift. The small patch of Brickearth lying in the bottom of the valley between Oldbury Hill and Crown Point, however, is a distinct deposit, probably of Head origin, and formerly extensively worked in pits 6 or 8 ft deep (Topley 1875, p. 192). Prestwich (1889, p. 287) records the finding of a large white Palaeolithic flake here at a depth of 4 ft. A small area of Brickearth [518606] on the east bank of the Darent about 1 mile N. of Otford is a heavy loam derived from local Head material on the chalk.   H.G.D., S.C.A.H.

The well-known fossiliferous fissures in Hythe Beds near Ightham [60655660] were discovered by Benjamin Harrison in 1891 and 1892. The site, then known as Pink's Quarry and now absorbed into the Borough Green quarries of the British Quarrying Co. Ltd., was on the east side of the lane from Borough Green to Basted, between Basted House and Basted Mill; the topography has been altered by recent tipping and quarrying and the original fissures have been quarried away.

The heterogeneous material filling the fissures included loam and brickearth, perhaps of Head origin, though possibly including some water-deposited material. The fauna is listed on pp. 140-2. The deposits were mixed and disturbed and abundant large and small fragments of Hythe Beds 'rag' occurred in the matrix. According to manuscript notes by the late A. S. Kennard, there were at least four fissures, of which three were filled with Pleistocene sandy brickearth and yielded most of the larger bones and a few of the smaller ones, mainly birds; the fourth, 'main' fissure was open to the sky at first, but sealed later, and most of the smaller bones and the shells and plants came from a friable earth with carbonaceous layers towards the top of this, though some of the larger Pleistocene species also occurred in it.

Mr. J. N. Carreck has kindly provided the following notes and the faunal list: "The problem is to distinguish between the remains of Pleistocene species, most of which fell or were washed into the fissures (though others, such as the bats and birds of prey, lived in them) from post-Glacial species, especially as there is seldom any difference in the condition of preservation of the bones. It seems likely that Abbott (1894), Newton (1894) and Howorth (*in* Newton 1894) were right in concluding that post-Glacial contamination was practically confined to that caused by burrowing and predatory animals. The main mammalian assemblage at least is all of one age and the filling of the fissures probably took place comparatively quickly. The fauna includes about equal numbers of arctic and subarctic species characteristic of tundra and loess-steppe environments and of species indicating a northern-temperate climate like that of northern Scotland or southern Norway today. An interstadial phase of no great severity is suggested and the truly Pleistocene species point to a time early in the last (Weichselian) glaciation. The microtine fauna can be correlated with that of other localities in Kent, Devon, Somerset and Derbyshire and the assemblage is found only with Late Palaeolithic flint implements. The bird fauna is not inconsistent with the above interpretation, for only six species do not range farther north than the north-temperate zone and some breed today as high as 70° N. On the other hand, the Mollusca are a mixture of species of several ages and ecological environments. The alpine species *Discus ruderatus* was probably extinct in Kent by about 5000 B.C., but *Helix aspersa* is unknown before Roman times at the earliest.

The fossils were described by Abbott (1894; 1907a), Bell (1915), Hinton (1910; 1926), Kennard and Woodward (1901), Newton (1894; 1899) and Reynolds (1909; 1912). They are briefly discussed by Jackson (*in* Cullingford 1953, pp. 206, 241).

Fossils from the Ightham Fissures

Plants

Chara sp.	Stonewort
Eurhynchium praelongum Hobkirk	Moss
Corylus avellana Linné	Hazel
Quercus robur Linné	Oak

Mollusca

Arianta arbustorum (Linné), *Carychium minimum* Müller, *C. tridentatum* (Risso), *Catinella arenaria* (Bouchard-Chantereaux), *Cecilioides acicula* (Müller), *Cepaea hortensis* (Müller), *C. nemoralis* (Linné), *Clausilia bidentata* (Ström), *Cochlicopa lubrica* (Müller) (aggregate), *Discus rotundatus* (Müller), *D. ruderatus* (Férussac), *Euconulus fulvus* (Müller), *Helicigona lapicida* (Linné), *Helicella itala* (Linné), *Helix aspersa* Müller, *Hygromia hispida* (Linné), *H. striolata* (C. Pfeiffer), '*Limax*' *spp.*, *Marpessa laminata* (Montagu), *Oxychilus alliarius* (Müller), *O. cellarius* (Müller), *Pomatias elegans* (Müller), *Pupilla muscorum* (Linné), *Retinella nitidula* (Draparnaud), *R. radiatula* (Alder), *Succinea oblonga* (Draparnaud), *Truncatellina cylindrica* (Férussac), *Vallonia costata* (Müller), *V. excentrica* Sterki, *Vitrea crystallina* (Müller) (aggregate).

Ostracoda
Candona candida Müller

Isopoda

Porcellio scaber auctorum	Woodlouse

Myriapoda

Julus sp.	Millipede

Insecta
Cynips sp. (galls)   Gall Wasp
Chrysomela sp.   Leaf Beetle
Otiorhynchus sp.   Curculionid Weevil

Amphibia
Bufo bufo (Linné)   Common Toad
Rana temporaria Linné   Common Frog
Triturus sp.   Newt

Reptilia
Anguis fragilis Linné   Slow-worm
Natrix natrix (Linné)   Grass Snake
Vipera berus (Linné)   Viper

Aves
Alauda arvensis Linné   Skylark
Anser cf. *anser* (Linné)   cf. Grey-lag Goose
A. cf. *platyrhyncha* Linné   cf. Mallard
A. sp.   Duck
cf. *Anthus pratensis* (Linné)   cf. Meadow Pipit
Aquila? sp.   Eagle
Buteo sp.   ?Buzzard
Falco peregrinus Tunstall   Peregrine Falcon
cf. *Fringilla coelebs* Linné   cf. Chaffinch
Hirundo rustica Linné   Swallow
cf. *Lanius collurio* Linné   cf. Red-backed Shrike
Larus? or *Sterna?*   Gull or Tern
Melanitta nigra (Linné)   Common Scoter
Motacilla cf. *alba* Linné
    yarrellii Gould   cf. Pied Wagtail
cf. *Oenanthe oenanthe* (Linné)   cf. Wheatear
Passer domesticus (Linné)   House Sparrow
cf. *Prunella modularis* (Linné)   cf. Hedge Sparrow
Rallus aquaticus Linné   Water Rail
Spatula clypeata (Linné)   Shoveller
cf. *Turdus ericetorum* Turton   cf. Song Thrush
cf. *T. merula* Linné   cf. Blackbird

Mammalia
  Artiodactyla
Capreolus capreolus (Linné)   Roe Deer
Cervus elaphus Linné   Red Deer
Ovibos moschatus (Zimmermann)   Musk Ox
Rangifer tarandus (Linné)   Reindeer
Sus scrofa Linné   Wild Pig

  Perissodactyla
Coelodonta antiquitatis (Blumenbach)   Woolly Rhinoceros
Equus caballus Linné   Wild Horse

  Proboscidea
Mammuthus cf. *primigenius* (Blumenbach)   cf. Mammoth

  Rodentia
Apodemus lewisi (Newton)   Field Mouse
A. sylvaticus (Linné)   Common Field Mouse
Arvicola abbotti Hinton   Water Vole
Citellus erythrogenoides (Falconer)   Suslik
Clethrionomys harrisoni (Hinton)   Bank Vole

Rodentia

C. kennardi (Hinton)	Bank Vole
Dicrostonyx henseli Hinton	Banded Lemming
Lemmus lemmus (Linné)	Norwegian Lemming
Microtus agrestis (Linné)	Common Field Vole
M. anglicus Hinton	Vole
M. arvalis (Pallas)	European Field Vole
M. corneri Hinton	Vole
M. ratticeps Keyserling and Blasius	Northern Vole

Lagomorpha

Lepus timidus Linné	
anglicus Hinton	Giant Hare
Ochotona spelaea Owen	Cave Pika

Carnivora

Alopex lagopus (Linné)	Arctic Fox
Canis lupus Linné	Wolf
'*Hyaena*' cf. *crocuta* Erxleben	
spelaea (Goldfuss)	cf. Cave Hyaena
? Lutra lutra (Linné)	? Otter
Meles meles (Linné)	Badger
Mustela nivalis Linné	Common Weasel
M. nivalis minuta Newton	Pigmy Weasel
M. putorius Linné	Common Polecat
M. robusta (Newton)	Giant Polecat
Ursus cf. *arctos* Linné	cf. Brown Bear
Vulpes vulpes (Linné)	Common Fox

Chiroptera

Myotis cf. *bechsteinii* (Kuhl)	cf. Bechstein's Bat
M. cf. *daubentonii* (Kuhl)	cf. Daubenton's Bat
(or *? Vespertilio murinus* Linné)	(or ? Parti-coloured Bat)
M. nattereri (Kuhl)	Natterer's Bat
?Pipistrellus pipistrellus (Schreber)	? Pipistrelle
Plecotus auritus (Linné)	Long-eared Bat

'*Vespertilio*' *sp.*, intermediate in size between Pipistrelle and Long-eared Bat

Insectivora

Neomys cf. *fodiens* (Schreber)	cf. Water Shrew
Sorex araneus Linné	Common Shrew
S. minutus Linné	Pigmy Shrew
Talpa europaea Linné	Common Mole

Anthropoidea

Homo sapiens Linné	Flint flakes"

Further fissures have been revealed in recent working back of the quarry faces. Some of them appear as 'gulls' filled to the surface with brickearth and stony loam, the materials clearly having been essentially introduced from above rather than introduced by waters circulating in open fissures. The fissures originated by cambering (see Worssam 1963, pp. 127–9) perhaps aided or modified by solution (cf. Topley 1894, p. 210).

Around New Town and north-west of West Malling a flat-lying area of Hythe Beds is covered by a thin mantle of fine buff or brown loam or Brickearth; it gives rise to a distinctive soil but is too irregularly distributed or thin for representation on the map. At the top of an old quarry [67835830] ½ mile N. of St. Mary's Church the drift was seen to be about 3 ft thick and to contain lumps of ragstone and chert as well as some flints. A patch of similar stony Brickearth shown on the map [678564]

¾ mile S. of West Malling church was formerly worked to 4 ft and more for brick-making. It is sandy and includes material probably derived from the adjacent Sandgate Beds. S.C.A.H.

LANDSLIP

The slips on the Hastings Beds are small. The largest [574438], measuring 200 x 300 yd moved in 1936 (see Survey Photographs A 7090–2). Smaller ones 300 yd E. of Great Bounds [57654335] and 250 yd N.E. of Mabledon Farm [58754470] are easily recognizable by their irregular surfaces and, at the former locality, by the lines of sedge growing along numerous spring lines. A slip of Wadhurst Clay in the grounds of Mabledon Hospital [58004445] occurred in March 1967.

Large slips on the Lower Greensand/Weald Clay junction are shown on the 1-inch map 1¼ miles N.N.W. of Shipborne, 2 miles S. of Westerham and 1½ miles S.E. of West Peckham. Others, recognized by the irregular, hummocky ground and associated seepages, but not mapped separately, can be traced continuously between the two western slips.

A slide of Weald Clay in the railway cutting [53555100] at the southern entrance to the Sevenoaks railway tunnel is described and figured by Terzaghi (1950, fig. 3, pl. 2, p. 97). The movement took place in 1939, about seventy years after the cut had been excavated, and was due to the decrease of cohesion produced by direct action of weathering. No springs or other indications of seepage could be observed.

At 300 yd N. of Puddledock a slipped mass of sand with blocks of hassock was noted [46155125]. A 5-ft slip of Atherfield and Weald Clay took place in January 1955 250 yd S. of Bardogs [468509]. 'Tumbled' hassock occurs in the road [47075111] leading up Toy's Hill. An old pit [48145160] opened in disturbed ground 500 yd W. of Ide Hill church revealed chert and rag rubble with some blocks up to 6 ft across. At 650 yd S.E. of Ide Hill church a slipped mass of chert and rag in a yellow loamy sand forms a small conical outlier [49055130] of Hythe Beds. Disturbed chert rubble was again noted in a pit [49595157] 700 yd E. of Hanging Bank. Cambering of the Hythe Beds could be clearly seen at Bayley's Hill [51525181] and 300 yd E. an old pit [51855192] had been opened in a slipped mass of rag. Many of the oak trees in the River Hill area show adjustment to verticality and because of internal cracks, due to continual straining, have proved useless for timber. There are no oaks older than 150 years. 'Tumbled' rag occurs on River Hill [54125210]. The field [547519] 600 yd E.S.E. of Riverhill House is stated by Colonel Rogers to have developed its hummocky surface between 1916 and 1932. Small slips [56045293] in 1936, 600 yd N.N.W. of Absalom's Farm exposed yellow and grey mottled sandy clay resting on black shaly clay with small nodules. Sections opened at Hubbard's Hill following the large scale slipping in April 1965 on the line of the Sevenoaks By-pass confirmed the observations, made by Dines in 1932, of the slipped nature of the ground. At one point [53475200] 15 ft of sheared and disturbed Weald Clay with blocks of Hythe Beds material at the base rested on 8 ft of roughly bedded sandy loam with coarse sand and debris from the Hythe Beds and at its base a heavier sandy clay loam with stones, resting on Weald Clay.

The investigations (Weeks in press) have shown that the Hythe Beds are both cambered and landslipped in a southerly direction. The cambering extends for at least 600 yd north from the escarpment.

Landslipping was found to have taken place at the edge of the escarpment and also at the southern edge of cambered blocks, 100 yd back from the escarpment face. The first activity involves the Hythe Beds slipping on the underlying Atherfield Clay; whilst the second involves the Hythe Beds only, the vertical south face of a cambered block having slipped into a fissure space between two cambered blocks.

The underlying Weald and Atherfield Clays are valley-bulged in front of the escarpment and are covered by a widespread solifluction sheet (ascribed to the Main Weichselian) that extends more than ¾ mile from the escarpment. In some areas this sheet is overlain by more recent and topographically evident solifluction lobes. Radiocarbon dating of a fossil soil found on the interface between a 14-ft thick recent lobe and the earlier sheet gave an age of 12 200 B.P., which corresponds to the Allerød Interstadial. This suggests that the lobes were formed initially during Zone III (Skempton and Petley 1967).

Remnants of an older chert-laden solifluction sheet are found capping the hills to the south of this area and these are ascribed to the Saale Glaciation (Gipping).

Slip surfaces were found on the interface between the recent lobes and the underlying sheet. Similarly they were present also between the underlying sheet and the Weald Clay, on which this sheet rests. In addition the top layers of the Weald Clay, which were found to be cryoturbated, also contained slip surfaces to a depth of 11 ft, and these layers had, therefore, become part of the soliflucted mass.

This last phenomenon has also been recorded on other slopes of Weald Clay across the Weald and again on slopes of Wadhurst Clay near Tonbridge and Gault at Wrotham Heath. A trial pit at Wrotham Heath showed 6 ft of flint and rounded chalk fragments overlying Gault clay, which was slickensided to a depth of 7 ft 6 in and weathered to a depth of 10 ft (Weeks *in press*).   C.R.B., S.C.A.H.

Chapter VII

ECONOMIC GEOLOGY

In the Sevenoaks area the more important aspects of economic geology are the provision of water supplies from underground sources; the digging of clay, sand, gravel and quarried stone for road metal and building purposes; and applications to agriculture, particularly regarding land drainage. As a matter of historical significance, the former winning, smelting or forging of iron ore from the Wealden Beds was carried out at fourteen or more sites within the area of the map.

CLAY FOR BRICKMAKING AND CEMENT MANUFACTURE

Numerous brickworks have been established from time to time on the clay formations of the Wadhurst Clay, the Weald Clay and the Gault; the Wadhurst Clay and the Weald Clay are particularly suitable for making bricks for a wide range of purposes. Some of these brickworks were small, and many are now closed. Wrotham Old Pottery, ⅓ mile N. of Borough Green Station, later became a brickfield. Present day works generally are on a large scale; the following is a list of locations:

(a) Working Wadhurst Clay and Lower Tunbridge Wells Sand. Quarry Hill, Tonbridge; High Brooms, Southborough.

(b) Working Weald Clay. Crowhurst; Holland, Limpsfield.

(c) Working Gault. Greatness, Sevenoaks; Dunton Green.

(d) Working Gault for cement manufacture. Wrotham; Ford Place, near Wrotham.

Although the future existence of the Quarry Hill pit [585450] has been threatened by the proposed Tonbridge By-pass (Anon. 1967) the High Brooms Brick and Tile Co. Ltd. thrives. When started as a company in 1885 the main products were commons made by the stiff plastic process. Later they changed to the wirecut process. Their hand-made trade commenced in 1900 and now numbers 5 million bricks per year. Partial mechanization in the form of a mechanical stoker for the kilns and delivery of clay to the hand moulder was undertaken in 1947. In 1963 an automatic processing plant was installed, and it is hoped that soon the company will have achieved a fully automatic brick factory. Formerly both Lower Tunbridge Wells Sand and Wadhurst Clay were mixed at the bottom of the pit as dug, but as a result of the modernization, each material is dug separately and fed into a box feeder. The sand and sandstone is reduced to minus ¾ inch in the hammermill and then elevated to hoppers. The feeding arrangements for the clay are similar. Different blends of sand and clay are required for the wirecuts (1 to 1) and the handmades (2 sand to 1 clay) and this is carried out automatically (Anon. 1963).

In the past, both brickearths of a type included with Head deposits and loamy parts of the Clay-with-flints have been dug for brickmaking on a small scale, the former, for example, eastward of Borough Green and the latter $\frac{1}{2}$ mile S.E. of Costains Farm (Brickyard Farm, 427591).

CHALK

Middle Chalk is at present worked, mainly for lime burning, north-west of Brasted and above Dunton Green. Numerous disused chalk pits bear testimony to past production, mostly for local use in liming the soil.

SAND AND GRAVEL

Sand. The Folkestone Beds of the Lower Greensand are the source of much of the building sand used in South-east England. Very extensive workings occur at intervals along the whole length of the outcrop of the Folkestone Beds, and worked-out areas already cover many acres. Numbers of smaller pits have exhausted their immediate reserves and fallen into disuse. In general, coarse, often ironstained sand occurs in the uppermost part of the Folkestone Beds, and in a number of sandpits several feet of Gault are removed as overburden. This coarse sand complies with British Standards Specifications 1198–1200 for building and plastering sand. Locally, as near Riverhead, sand may be of sufficiently coarse grade to comply with B.S.S.882 for fine aggregate for concreting.

Below the generally coarse sand the grade is finer and the sand is commonly white. This is quarried as a 'silica sand' for glass manufacture, moulding sand and other special purposes. The simple arrangement of coarse sand above and finer sands below, however, does not persist everywhere along the outcrop; nor does it extend to the lower beds. Some of the lowest sand, as at Wrotham Heath, is very variable and highly impregnated with iron in stony layers and concretions. There are also a loamy facies, bands of silt and lenticular sandrock.

Sand from the Folkestone Beds is used in the manufacture of sand-lime bricks at Sevenoaks, Ightham and Ryarsh. At Sevenoaks (Greatness) the pits are flooded and the sand is extracted hydraulically.

So extensively has sand been worked that available reserves are limited for a long-term policy of quarrying, although alteration of user of certain tracts would probably release further supplies.

The following is the general location of quarries working the Folkestone Beds in 1955. They are listed from west to east along the outcrop.

Westerham; Brasted; Chipstead, near Riverhead; Riverhead; Bradbourne, near Riverhead; Greatness, near Sevenoaks; Otford; Stone Street; Ightham; Borough Green; Ford Place, near Wrotham; Addington; Ryarsh.

A fine white to pale yellow soft sandrock, roofed by a more siliceous band of sandstone, was formerly worked in chambers near Combe Bank, Sundridge; at Chipstead Whitening Works similar soft white sandrock was excavated in chambers some 20 ft high.

Sharp sand is extracted with gravel from below the alluvium of the rivers Medway and Darent, notably at East Peckham, Sundridge and Riverhead. At the last two places it includes sand from the Folkestone Beds obtained hydraulically by suction to 50 ft depth, the materials being collected in a

(A. 10282)

A.   SANDPIT IN FOLKESTONE BEDS OVERLAIN BY GAULT CLAY;
NORTH OF COVER'S FARM, WESTERHAM

PLATE X

B.   SAND DREDGING OF FOLKESTONE BEDS, RIVERHEAD, NEAR SEVENOAKS

(A. 9920)

cruising barge in which fine silt flows over and coarse gravel, with sand on top, settles to the bottom.

Soft, fine sand was formerly dug from the Tunbridge Wells Sand near Pembury, and from various sites on the Ashdown Beds.

Gravel. River Gravels are exploited at Sundridge, Riverhead and East Peckham, either by dry excavation of terraced gravels or dredged from beneath the Alluvium.

BUILDING STONE AND ROAD METAL

These categories of stone are here considered together since the bulk of both materials is quarried from the Hythe Beds. In the past this formation has yielded much Kentish Rag for building stone, but little is used for that purpose today. On the other hand, it is now very extensively quarried for road metal, chiefly at Borough Green and at Offham, whence it has also supplied rockery and sea-walling stone. Sabine (*in* Worssam 1963, p. 135) gives the results of mechanical tests upon Kentish Ragstone from the Borough Green and Seal Hollow quarries within this area and also from two quarries which are within the area of the Maidstone (288) Sheet.

Chert, either from the Angular Chert Drift or directly from seams ('Sevenoaks Stone') in the Hythe Beds, was formerly largely used for road metal, but today its extraction appears to be restricted to a single quarry (St. Julian's) about 2 miles S.S.E. of Sevenoaks. The larger blocks of chert, cherty sandstone, nodular sandstone and ragstone from the Hythe Beds have been dug for road foundations from quarries at The Chart, Limpsfield, and at Borough Green and elsewhere; much stone of smaller grade was obtained, however, from the many old shallow workings in the Angular Chert Drift by screening off the sandy matrix. Ightham Stone, of the Folkestone Beds, was worked early in the nineteenth century for a supply of road metal to London; Oldbury Stone was also formerly used.

In former years flints from the Chalk, hard chalk, and ironstone from the Folkestone Beds were all used as local building materials. Calcareous tufa, commonly used in Norman buildings on account of its ease in cutting and squaring before hardening, occurs at Basted, near Ightham.

On the Chalk outcrop flints from stony drift deposits continue to provide useful local road metal, gathered from fields or shallow pits; the greater toughness of the white varieties also increases with exposure to the atmosphere. In the past, other sources of road metal have included River Gravels, flinty Head, various Wealden rock-bands and the slag or 'cinder' from old ironworks. The ferruginous carstone locally concentrated in the Folkestone Beds has been used in Limpsfield and Westerham, for instance, for paving; the stone is set on end with the bedding planes vertical.   F.H.E., H.G.D., S.C.A.H., C.R.B.

AGRICULTURE

The geology of a region is one of several factors that control the agricultural use of land. Its chief influence is on soil type, on natural drainage, and on contour of the ground. At one period, clay lands of the Wadhurst Clay, the Weald Clay and the Gault were given over largely to pasture. Over large tracts the surface soil is mixed, but drainage was impeded by the heavy clay below. In many districts these clay lands have now been artificially drained and large areas formerly of pasture are now under the plough.

On the whole the soils of the Lower Greensand outcrop are less suitable for arable cultivation than those of other formations. Drainage tends to be excessive, and much of the Lower Greensand country is rough pasture or woodland. The word 'Chart', indicating a dry heath, appears on the map in a number of places. It is thought that Mereworth Woods may possibly be a remnant of the permanent woodland of the Weald.

Where an ample supply of water for irrigation is available, parts of the Lower Greensand can make admirable market garden and nursery land.

Much of the Chalk outcrop is drift covered and as in the case of the clay lands many acres of former pasture are today arable. The chief remaining pasture is on the steep slopes of the escarpment and of the sides of the dry valleys.   F.H.E.

WATER SUPPLY

The Sevenoaks district is almost entirely dependent on underground sources of water. A small surface reservoir is present at Pembury Waterworks [627426] but is fed from springs rising from the base of the Lower Tunbridge Wells Sand and from boreholes penetrating the Ashdown Beds. Previous published information on the hydrogeology of the area is contained in Whitaker (1908, 1912), Buchan and others (1940), and Cooling and others (1968).

In the past, attempts were made to utilize each of the formations present in the district for water supply purposes but in recent years attention has been concentrated on the two main aquifers, the Lower Greensand and the Hastings Beds. Table 2 shows the annual abstractions (million gallons) from these aquifers and from the next most important aquifer, the Chalk, for the Sevenoaks district for the year ended 31st October 1963. This information has been abstracted from the annual returns made under Section 6 of the Water Act 1945. Additional abstraction is estimated at 5 per cent of the recorded abstraction.

TABLE 2

Abstraction in millions of gallons from the
principal aquifers of the Sevenoaks district
for the year ended 31st October 1963

Aquifer	Public Supply	Others	Total
Chalk	73	—	73
Lower Greensand (undivided)	1937	—	1937
Folkestone Beds	—	4	4
Hythe Beds	588	—	588
Total: Lower Greensand	2525	4	2529
Hastings Beds (undivided)	333	—	333
Lower Tunbridge Wells Sand	—	8	8
Ashdown Beds	477	10	487
Total: Hastings Beds	810	18	828
Grand totals:	3408	22	3430

The chemical quality of ground water from the principal aquifers is generally satisfactory and the water is suitable for a wide range of industrial and other purposes. Typical chemical anlyses are given in Table 3.

Drift. Where patches of permeable sands and gravels overlie an impermeable solid substratum, for example either Head on Sandgate Beds, or Head, Brickearth, or River Gravels on Weald Clay, small amounts of ground water accumulate which in the past were used for domestic supply. However, these sources are liable to pollution and marked seasonal variation in yield and hence are now rarely utilized.

Alluvial sands and gravels flanking the more important rivers of the district may contain appreciable quantities of ground water. At New Wharf Road, Tonbridge [588464], up to 500 000 gallons per day (gal/day) are obtained from a network of 9-in drains set in 10 ft of river gravels. The bacterial and chemical quality of the water is good, the total hardness being 170 milligrams per litre (mg/l) and the total dissolved solids 240 mg/l. The iron content is low (0·24 mg/l).

Chalk and Upper Greensand. As the Upper Greensand, where present, is in hydraulic continuity with the Chalk, the two formations can be considered together. However, it must be noted that whereas in the Upper Greensand most ground-water flow takes place in the intergranular pore space, in the Chalk, water movement occurs predominantly along fissures, joint and fault planes and lines of flint.

Over the greater part of the Chalk scarp within the Sevenoaks district. the Upper Greensand is absent and springs rise at the junction of the Chalk with the underlying impermeable Gault. These may in some cases have appreciable flows and Whitaker (1908) records 1 140 000 gal/day from a spring near Court Lodge, Brasted [462561]. In the west, Chalk ground water passes downwards through the Upper Greensand and emerges at springs slightly lower in the scarp. None of the springs is utilized as a source for public supply. On the dip-slope of the Chalk, the valleys within the sheet boundary are dry and no springs occur.

The Chalk in the Sevenoaks district does not constitute a major source of ground water for use within the district but acts as part of the catchment area for supplies taken from the formation in the districts to the north. The water table, though conforming in a general way to the ground surface morphology, is at considerable depth below ground level over much of the south of the outcrop and may be as low as 400 ft below some sections of the crest of the North Downs. Consequently the horizons within the zone of permanent saturation are those of the marly Lower Chalk in which flow is restricted by the relatively low degree of fissuring and by the absence of flint bands. At only one point on the Chalk scarp are these horizons tapped for public supply. At Westerham Hill Pumping Station [428558] two linked shafts, each with continuing bores passing through the underlying Upper Greensand to the Gault, yield 9000 gallons per hour (gal/hr).

To the north the water table rises through the succession and in the north-west of the district it extends above the hard, fractured Melbourn Rock at the base of the Middle Chalk. This horizon and the underlying strata have been developed in the Jewels Wood Borehole [406607] where a 30-in diameter bore reducing to 24-in at depth, passing from a horizon some 160 ft above the base of the Melbourn Rock to bottom the Upper Greensand yields 5000 gal/hr.

The chemical quality of the water from these two wells corresponds with that obtained from these horizons in contiguous areas. Total hardness ranges from 197 to 250 mg/l at Jewels Wood and from 251 to 280 mg/l at Westerham Hill, the corresponding non-carbonate hardness ranges being 8 to 67 mg/l and 32 to 60 mg/l respectively.

Gault. The hydrogeological importance of the Gault lies in its impermeable nature. Although in the past ground water in some of the more silty beds was very occasionally used for domestic supply, the formation essentially functions as a seal to ground water in the underlying Lower Greensand. Thus not only is the Lower Greensand ground water confined under pressure, but it is protected from infiltration by polluted surface waters. It is for this reason that many modern wells extracting from the Lower Greensand have been sited on the Gault. An additional benefit from such sites is a decrease in the hardness of the water as a result of cation exchange reactions occurring beneath the Gault sediments (see below).

Lower Greensand. Both the Folkestone Beds and the Hythe Beds function as aquifers, the latter being the more important. The Sandgate Beds, although thin, appear to act as an impermeable seal separating ground waters in these two permeable formations over the whole district. The Atherfield Clay is also impermeable.

The presence of saddle folds in the Lower Greensand outcrop (see page 11 and Fig. 5) may locally affect the hydrogeological behaviour of the aquifers. In unconfined conditions these folds can have a ponding effect on ground water up dip of the fold and may also affect the fissuring of the more compact beds. In assessing the hydrogeological characteristics of any site in this area, attention must therefore be paid to the local detail of the geological structure as well as to the nature of the aquifers themselves.

In the sands of the Folkestone Beds ground-water movement takes place by intergranular flow and yields are much affected by the total thickness of aquifer. In the west of the district where thicknesses are at a maximum (see page 59), a 15-in bore at Limpsfield [399541] yielded 40 000 gal/hr over a three-day test period, during which the drawdown from the rest water level was only 30 ft. In the central part of the district yields are somewhat less, although still considerable. A 40-in bore reducing to 36-in at Brasted [470557] yielded 36 000 gal/hr over a 14-day test, although here the drawdown was 104 ft. In the east the Folkestone Beds are developed at two major abstraction points, but as they are pumped jointly with the Hythe Beds only a combined yield is available (see below).

Over the whole of the district the unconsolidated and relatively fine-grained nature of much of the Folkestone Beds necessitates the use in boreholes of gravel packs or sand screens to avoid the problems of excessive sand contamination at high pumping rates.

Springs are not frequent on the Folkestone Beds outcrop but may rise at the base of the formation where they are thrown out by the impermeable Sandgate Beds, within the formation where they result from the presence of thin clay layers, or on the dip-slope at the contact with the overlying Gault.

Ground water from the Folkestone Beds at outcrop is moderately hard with values of total hardness around 180 mg/l and of non-carbonate hardness around 80 mg/l. Under Gault cover, however, hardness may be reduced, values for total hardness of around 110 mg/l and for non-carbonate hardness of around 50 mg/l being recorded. It seems likely that this is the result of a natural cation exchange reaction of the type referred to above. The iron content is generally low at less than 1 mg/l.

In the Hythe Beds thicknesses vary considerably and there is an important lithological change across the district from an arenaceous sequence in the west to a more calcareous 'rag and hassock' sequence in the centre and east (see page 57). This change has a marked effect on the characteristics of the aquifer.

In the sandy succession of the west intergranular flow is dominant and only moderate yields are obtained. At Westwood Pumping Station [424541] two 18-in bores and one 22-in bore have a combined yield of 27 000 gal/hr. Further to the east fissure flow in the cemented ragstone beds becomes important and yields increase markedly. At Sundridge Pumping Station [490557] two 24-in bores and one 16-in bore were test pumped for 28 days to give a combined yield of 134 000 gal/hr. The results of this test pumping were analysed by the Water Department of the Institute of Geological Sciences and a mean value of 156 000 gal/day/foot was obtained for the transmissibility of the Hythe Beds immediately adjacent to the pumped well. In the direction of Sevenoaks this value increased to 242 500 gal/day/foot.

In the region immediately to the north of Sevenoaks, ground water in the Hythe Beds is confined beneath the Sandgate Beds under artesian head, and at Cramptons Road Pumping Station [531569] a 24-in diameter bore overflowed at 96 000 gal/hr over an initial 11-day test period. In the east of the area Hythe Beds are pumped jointly with the Folkestone Beds and the combined yields are discussed below.

Springs are common at the base of the Hythe Beds and prior to 1915 were utilized for public supply at Bolthurst Farm, Paines Hill [413516]. Subsequently these springs were replaced by four shallow shafts, from which a yield of 10 000 gal/hr is obtained at present. Springs also rise at the junction of the Hythe Beds with the overlying Sandgate Beds in areas where the water table is sufficiently high. This type of spring is common around West Malling and in the central region where a spring at Sundridge [488554] yielded 11 000 gal/day until affected by pumping at the nearby Sundridge Pumping Station.

The hardness of the ground water from the Hythe Beds shows a marked variation across the outcrop. In the west it is slightly hard with values of total hardness of 105 to 153 mg/l and of non-carbonate hardness from 15 to 39 mg/l. In the central region around Sevenoaks it is hard, total hardness ranging from 207 to 245 mg/l, and non-carbonate hardness from 36 to 67 mg/l. Over the whole area iron is generally low but may reach 3 mg/l at some localities.

The presence of relatively fine-grained incoherent sediments, both in the sandy successions in the west and in the 'hassock' beds further east, creates the risk of sand contamination at high pumping rates. This is generally overcome by the use of sand screens or gravel packs.

Where water is obtained from both the Folkestone Beds and the Hythe Beds, yields of boreholes are generally improved. In the west a 12-in bore at

South Green, Titsey [410542] yielded 20 000 gal/hr. In the east of the area two major abstractions take place. At Trosley Pumping Station, Trottiscliffe [640591] a 24-in bore yielded 55 700 gal/hr on 7-day test with a drawdown of only 32 ft. At Ryarsh Pumping Station [667605] on the sheet boundary a 36-in bore and a 24-in bore reducing to 18-in, jointly yield 40 000 gal/hr.

As might be expected, the chemistry of these mixed ground waters is intermediate between those of the separate aquifers. In the west the water is moderately soft and total hardness values of around 50 mg/l with corresponding non-carbonate values of around 35 mg/l have been recorded. Iron (up to 0·05 mg/l) and total solids (up to 132 mg/l) are low. In the east the water is hard with total hardnesses of from 198 to 225 mg/l, and non-carbonate hardnesses of 47 to 49 mg/l. Iron is low (0·7 to 0·9 mg/l) and total dissolved solids moderate (234 to 280 mg/l).

Weald Clay. This is essentially an impermeable clay formation but it also contains thin, silty sandstones and limestones which have been developed for local supply. Yields are generally low, the largest recorded being 420 gal/hr from a 3-in bore at Hadlow [637483]. Springs are occasionally associated with the outcrops of these more permeable beds and these too have been tapped for domestic and agricultural use in the past.

Hastings Beds. Over the greater part of the south of the Sevenoaks district, the Hastings Beds comprise three arenaceous aquifers, the Upper and Lower Tunbridge Wells Sands and the Ashdown Beds, separated by two predominantly argillaceous aquicludes, the Grinstead Clay and the Wadhurst Clay. To the east of Tonbridge the Grinstead Clay is not present, except within a single outlier and the Tunbridge Wells Sand is undivided.

Each of the aquifers contains a variety of lithologies which range in grain size from medium-grained sands to silty clays and in degree of cementation from hard sandstone to poorly consolidated sand; hence flows vary markedly in type, both intergranular and fissure flow occurring, and in rate. Aquifer thicknesses also vary and structural complexities arise as the result of the folding and faulting which have affected these beds. Consequently the forecasting of ground-water supply in these strata is fraught with considerable difficulty and the chances of success are rather less than in other, less complex areas.

Further difficulties arise from the presence of poorly consolidated horizons. If no preventive measures are taken, yields commonly diminish with time owing to the packing of fine-grained sediment in the intergranular pore space immediately adjacent to the well and to the silting of the well itself. In addition sand in suspension may create problems in water treatment. However, with modern methods of well construction, such as the use of gravel packs or sand screens, and of well development, these problems can generally be overcome.

The uppermost aquifer, the Upper Tunbridge Wells Sand, was utilized in the past for a small number of local supplies but it is thought that these have now (1967) all fallen into disuse. Yields were generally low but at Lingfield [403436] a 7-in bore yielded 1200 gal/hr. No analyses of ground water from this aquifer are available but it was reported to be soft though liable to be ferruginous.

TABLE 3

Typical analyses of ground water from the principal aquifers (in mg/l)

	MCk, LCk and UGS	F[2]	F[3]	H[4]	H[5]	F and H	LTW	A	LTW and A
Location	Jewels Wood, Biggin Hill	Greatness, Sevenoaks	Brasted	Paines Hill Limpsfield	Cramptons Road, Sevenoaks	Trosley, Trottiscliffe	Pembury	Haysden	Tonbridge
I.G.S. Ref. No.	287/98	287/67	287/101c	287/3	287/29c	287/53c	287/21	287/31a	287/1b and c
National Grid Reference	406607	535572	470557	413516	531569	640591	613413	560448	588464
Analyst†	(1)	(1)	(1)	(2)	(3)	(4)	(5)	(3)	(3)
Date	1961[1]	29.11.54	31.5.67	26.2.58	27.1.58	2.3.65	14.12.36	6.12.33	1.5.32
Total Solid Residue (dried at 180°C)	—	—	—	166	280	270	—	270	510
Total hardness	250	174	112	107	220	204	127	100	170
Non-carbonate hardness	26	79	47	39	40	56	27	5	Nil
Alkalinity	—	95	—	68	180	148	56	—	280
pH	7·2	6·7	6·2	7·6	7·5	6·8	7·4	7·6	7·3
Free CO_2	—	27	92	3	16	55	2	—	20
Total Iron	—	—	0·02	0·02	Nil	0·35	0·1	0·05	0·23
Calcium as Ca	—	—	—	38·8	—	—	58·8	—	—
Magnesium as Mg	—	5·5	4	2·5	—	—	9·0	—	—
Sodium as Na	9·0	15·0	17	2·5	21	19·5	24·8	—	61
Chloride as Cl	—	—	46	20	—	—	13·7	2	—
Sulphate as SO_4	2·2	7·0	5·5	15·8	4·3	3·3	2·0	1·2	1·4
N_2 as Nitrate	—	—	—	3·9	Nil	Nil	—	Nil	<0·01
N_2 as Nitrite	—	—	—	Nil	—	—	—	—	—
Fluoride as F	—	—	0·06	0·10	—	—	—	—	—

* Aquifer: MCk = Middle Chalk; LCk = Lower Chalk; UGS = Upper Greensand; F = Folkestone Beds; H = Hythe Beds; LTW = Lower Tunbridge Wells Sand; A = Ashdown Beds.

† Analyst: (1) Metropolitan Water Board. (2) East Surrey Water Company. (3) Counties Public Health Laboratories, Ltd. (4) Mid Kent Water Company. (5) The Permutit Company, Ltd.

Notes: [1] Average for three samples taken during the year; [2] At outcrop; [3] Under Gault cover; [4] In west of district; [5] In centre of district.

Although the underlying aquiclude, the Grinstead Clay, contains a number of somewhat more permeable horizons of sandstone and limestone, including the relatively thick Cuckfield Stone in the Chiddingstone Hoath region, springs are rare and no use is made of the formation for water supply.

Yields from bores in the Lower Tunbridge Wells Sand are variable, values ranging from nil to around 3000 gal/hr. Examples of the more successful wells are a 10-in bore reducing to 6-in at Old Powder Mills, Leigh [569466] which yielded 3500 gal/hr with a drawdown of only 4 ft, and a 12-in bore at Tonbridge [589466] which yielded 3000 gal/hr with a drawdown of 13 ft.

Springs are common at the base of the Lower Tunbridge Wells Sand and several have been utilized for public supply. In the area between Matfield and Pembury Wood a total of 18 springs have been tapped and led into the reservoir at the Waterworks [628418]. In 1966 the total yield varied from 153 000 gal/hr in March to 24 000 gal/hr in October. Springs at Modest Corner, Southborough [570422], yield from 1000 to 9000 gal/hr while a 10-ft deep shaft and headings on the site of Tubbs Hole Springs, Penshurst [515416], yield up to 3200 gal/hr.

The few analyses made of ground water from the Lower Tunbridge Wells Sand show a marked variation. Total hardness ranges from 127 to 374 mg/l with non-carbonate hardness from 27 to 70 mg/l. The degree of mineralization may be high, values for total solids of up to 729 mg/l being recorded. The iron content is commonly recorded as being high, although that of the analysed samples ranged up to only 1 mg/l.

The underlying Wadhurst Clay, though essentially an aquiclude, contains thin sandstones and limestones from which supplies have been obtained. These have been generally small but at Tonbridge [597459] a 6-in bore yielded 1500 gal/hr with a drawdown of 35 ft. No analyses of these ground waters are available.

The Ashdown Beds constitute the major aquifer of the Hastings Beds, abstraction from them comprising over 95 per cent of the total. Nevertheless yields are variable and several bores have failed to obtain their expected supply. Large diameter bores have generally been the more successful. At Pembury Waterworks [626425] a 32-in bore reducing to 20-in yielded 48 000 gal/hr on initial test in 1958, though this had dropped to 28 000 gal/hr by 1961, while at Saint's Hill Pumping Station [523414] an 18-in bore reducing to 15-in yielded 39 500 gal/hr for a drawdown of 18 ft. At some sites, headings have been constructed in order to increase yields; at Bidborough [560448] a 10-ft diameter shaft with a total of 752 ft of headings and deepened with a 9-in bore yielded 10 000 gal/hr with a drawdown of 18 ft.

Springs are occasionally thrown out at the junction with the overlying Wadhurst Clay where structural conditions are favourable, notably north-east of Southborough and to the south of Hever. These, however, are generally small and have not been made use of for supply.

As with the rest of the formation the analyses available are very variable. Total hardness varies from 29 to 294 mg/l and non-carbonate hardness from 5 to 127 mg/l. Total solids are generally fairly high with values from 270 to 564 mg/l. The iron content may in some cases be sufficiently high to warrant treatment for public supply but no analyses of these iron-rich waters are available.

One major abstraction takes water from both the Ashdown Beds and the overlying Lower Tunbridge Wells Sand. This is at New Wharf Road Pumping Station, Tonbridge [588464] where two shafts with headings in the upper aquifer and continuing bores into the lower yield up to 24 000 gal/hr. An analysis showed that the water was moderately hard with a total hardness of 170 mg/l which is entirely temporary. Iron and manganese were small but significant (0·2 and 0·3 mg/l respectively) and total solid content was moderate at 510 mg/l.

A small number of bores have attempted to tap all three aquifers from sites on the Weald Clay outcrop but these have generally been unsuccessful, either due to insufficient yield or rapid siltation. An exception is that at Style Place, Hadlow [646490] which in 1890 overflowed to a height of 24 ft above ground level at a rate of 2400 gal/hr. Subsequent yields were, however, somewhat less. Analyses of the water showed it to be highly alkaline with a sodium carbonate content calculated at about 650 mg/l (Whitaker 1908) and it seems likely that this is the result of cation exchange reactions which took place under prolonged contact with the interbedded clay formations.   A.C.B., S.B.

REFERENCES

ABBOTT, W. J. LEWIS, 1893. Excursion to Basted and Ightham. *Proc. Geol. Assoc.*, **13**, 157–62.

—— 1894. The Ossiferous Fissures in the Valley of the Shode, near Ightham, Kent. *Quart. J. Geol. Soc.*, **50**, 171–87.

—— 1907a. The Ossiferous Fissures of the Valley of the Shode. *In* Bennett, F. J. *Ightham: The story of a Kentish Village and its Surroundings.*

—— 1907b. Excursion to Tonbridge. *Proc. Geol. Assoc.*, **20**, 97–100.

—— 1916. The Pliocene Deposits of the South-East of England. *Proc. Prehist. Soc.*, **2**, 175–94.

—— and NEWTON, E. T. 1893. Excursion to Basted and Ightham. *Proc. Geol. Assoc.*, **13**, 157–62.

ALLEN, P. 1949. Wealden Petrology: The Top Ashdown Pebble Bed and the Top Ashdown Sandstone. *Quart. J. Geol. Soc.*, **104** for 1948, 257–321.

—— 1954. Geology and Geography of the London–North Sea Uplands in Wealden Times. *Geol. Mag.*, **91**, 498–508.

—— 1955. Age of the Wealden in north-western Europe. *Geol. Mag.*, **92**, 265–81.

—— 1959. The Wealden Environment: Anglo-Paris Basin. *Phil. Trans. Roy. Soc.* (B), **242**, 283–346.

—— 1960. Strand-Line Pebbles in the mid-Hastings Beds and the Geology of the London Uplands. General Features. Jurassic Pebbles. *Proc. Geol. Assoc.*, **71**, 156–65.

—— 1961. Ditto. Carboniferous Pebbles. *Proc. Geol. Assoc.*, **72**, 271–85.

—— 1962. The Hastings Beds Deltas: Recent Progress and Easter Field Meeting Report. *Proc. Geol. Assoc.*, **73**, 219–43.

—— 1967. Origin of the Hastings Facies in North Western Europe. *Proc. Geol. Assoc.*, **78**, 27–105.

—— and KRUMBEIN, W. C. 1962. Secondary Trend Components in the Top Ashdown Pebble Bed: A Case History. *J. Geol.*, **70**, 507–38.

ANDERSON, F. W. 1940. Ostracod zones of the Wealden and Purbeck. *Adv. of Sci.*, **1**, 259.

—— 1967. Ostracods from the Weald Clay of England. *Bull. Geol. Surv. Gt. Brit.* No. 27, 237–69.

——, BAZLEY, R. A. and SHEPHARD-THORN, E. R. 1967. The Sedimentary and Faunal Sequence of the Wadhurst Clay (Wealden) in Boreholes at Wadhurst Park, Sussex. *Bull. Geol. Surv. Gt. Brit.*, No. 27, 171–235.

—— and HUGHES, N. F. 1964. The 'Wealden' of North-West Germany and its English equivalents. *Nature*, **201**, 907–8.

ANON., 1963. Automatic Processing at High Brooms. *Br. Claywkr.*, **72**, 375–8.

—— 1967. *Claycraft*, **40**, (4), 162.

ARKELL, W. J. 1931. A Monograph of British Corallian Lamellibranchs, Pt. 3. London (*Palaeontogr. Soc.*).

—— 1941. The gastropods of the Purbeck Beds. *Quart. J. Geol. Soc.*, **97**, 79–128.

BATHER, F. A. 1911. Notes on Crinoid plates from the Penshurst Boring. *Summ. Prog. geol. Surv. Gt. Brit.* for 1910, 78–9.

BELL, A. 1915. Pleistocene and later bird fauna of Great Britain and Ireland. *Zoologist* (4), **19**, 401–12.

BENNETT, F. J. 1907. *Ightham: the story of a Kentish village and its surroundings.* London.

—— 1908a. Solution-Subsidence Valleys and Swallow-Holes within the Hythe Beds area of West Malling and Maidstone. *Geogr. J.*, **32**, 277–88.

—— 1908b. Map showing the Drifts of the Ightham area. In 'Museum Notes'. *South-Eastern Naturalist*, **13**, p. lxv, pl. II.

—— and HARRISON, B., 1906. Excursion to Borough Green, etc. and Ightham. *Proc. Geol. Assoc.*, **19**, 460–4.

BERDINNER, H. C. 1936. A section at Biggin Hill Aerodrome. *Proc. Geol. Assoc.*, **47**, 15–21.

BERRY, F. G. 1961. Longitudinal Ripples in the Upper Tunbridge Wells Delta, Kent, and their probable mode of origin. *Proc. Geol. Assoc.*, **72**, 33–9.

BIRD, E. C. F. 1963. Denudation of the Weald Clay Vale in West Kent. *Proc. Geol. Assoc.*, **74**, 445–55.

BONNEY, T. G. 1888. Note on the Structure of the Ightham Stone. *Geol. Mag.*, (3), **5**, 297–300.

BRITISH STANDARD 812 : 1960. Sampling and testing of mineral aggregates, sands and fillers. London.

BROWN, E. E. S. 1928. Report of Excursion to Borough Green. *Proc. Geol. Assoc.*, **39**, 194–5.

—— 1937. Field Meeting at West Malling, Kent. *Proc. Geol. Assoc.*, **48**, 396–8.

—— 1941. The Folkestone Sands and Base of the Gault near Wrotham Heath, Kent. With Report of the Field Meeting to Wrotham Heath and Offham, Kent. *Proc. Geol. Assoc.*, **52**, 1–15.

—— and HIMUS, G. W. 1938. Field Meeting at Wateringbury and Mereworth. *Proc. Geol. Assoc.*, **49**, 55–7.

BROWN, H. J. W. 1924. Notes on the Geology of the District around Borough Green, Kent. *Proc. Geol. Assoc.*, **35**, 79–83.

—— 1925. Minor Structures in the Lower Greensand of W. Kent and E. Surrey. *Geol. Mag.*, **62**, 439–51.

BUCHAN, S. 1938. Notes on some outliers of Grinstead Clay around Tunbridge Wells, Kent. *Proc. Geol. Assoc.*, **49**, 407–9.

——, ROBBIE, J. A., HOLMES, S. C. A., EARP, J. R., BUNT, E. F. and MORRIS, L. S. O. 1940. Water Supply of South-East England from Underground Sources. *Geol. Surv.* Wartime Pamphlet (10) Pt. VI.

BURY, H. 1910. The Denudation of the Western End of the Weald. *Quart. J. Geol. Soc. Lond.*, **66**, 640–92.

CASEY, R. 1946. The Folkestone Beds: Aeolian or Marine? *South-Eastern Naturalist*, **51**, 43–7.

—— 1951. In 'Geological Records'. *South-Eastern Naturalist*, **55**, xxiv–xxv.

CASEY, R. 1954. *Falciferella*, a new genus of Gault ammonites, with a review of the family Aconeceratidae in the British Cretaceous. *Proc. Geol. Assoc.*, **65**, 262–77.

—— 1955. The pelecypod family Corbiculidae in the Mesozoic of Europe and the Near East. *J. Washington. Acad. Sci.* **45** (12), 366–72.

—— 1958. in *Summ. Prog. geol. Surv. Gt. Brit.* for 1957, 48.

—— 1959. Field Meeting at Wrotham and the Maidstone By-pass. *Proc. Geol. Assoc.*, **70**, 206–9.

—— 1960–1966. A Monograph of the Ammonoidea of the Lower Greensand. Pts. I–VII. London (*Palaeontogr. Soc.*).

—— 1961. The Stratigraphical Palaeontology of the Lower Greensand. *Palaeontology*, **3**, 487–621.

—— 1963. The Dawn of the Cretaceous Period in Britain. *Bull. S.-E. Un. sci. Socs.*, **117**.

CAYEUX, L. 1916. Introduction à l'étude pétrographique des roches sédimentaires. *Mém. pour servir à l'explic. carte géol. dét. France*.

CHALONER, W. G. 1962. Rhaeto-Liassic Plants from the Henfield Borehole. *Bull. Geol. Surv. Gt. Brit.*, **19**, 16–28.

CHANDLER, R. H. and LEACH, A. L. 1909. Excursion to Otford and the Darent Valley. *Proc. Geol. Assoc.*, **21**, 236–40.

—— —— 1932. The Eocene Outlier near Knockmill, Kent; and Report of Field Meeting. *Proc. Geol. Assoc.*, **43**, 284–9.

—— —— 1936. The Structure of the Eocene Outlier near Knockmill, Kent. *Proc. Geol. Assoc.*, **47**, 239–48.

COOLING, C. M. and others. 1968. Records of Wells in the Area of New Series One-Inch (Geological) Reigate (286) and Sevenoaks (287) sheets. *Water Supply Pap. Inst. Geol. Sci., Well. Cat. Ser.*

COPE, J. C. W. 1967. The Palaeontology and Stratigraphy of the lower part of the Upper Kimmeridge Clay of Dorset. *Bull. Brit. Mus. (Nat. Hist.) Geol.*, **15**, 1–79.

CULLINGFORD, C. H. D. 1953. *British Caving*. London.

DAVIES, G. M. 1914. *Geological Excursions around London*. London.

—— 1916. The rocks and minerals of the Croydon Regional Survey Area. *Proc. and Trans. Croydon Nat. Hist. Soc.*, **8**, pt. 2, 53–96.

DAVIS, A. G. 1926. Notes on some Chalk Sections in N.E. Surrey. *Proc. Geol. Assoc.*, **37**, 211–220.

DAWKINS, W. B. 1905. *In Royal Commission on Coal Supplies:* Final Report, pt. x, 26–41.

DEWEY, H., BROMEHEAD, C. E. N., CHATWIN, C. P. and DINES, H. G. 1924. The Geology of the Country around Dartford. *Mem. Geol. Surv.*

DIBLEY, G. E. 1900. Zonal Features of the Chalk Pits in the Rochester, Gravesend and Croydon Areas. *Proc. Geol. Assoc.*, **16**, 484–99.

—— 1918. Additional Notes on the Chalk of the Medway valley, Gravesend, West Kent, North-east Surrey and Grays (Essex). With Palaeontological Notes and Appendices by G. C. Crick and R. B. Newton. *Proc. Geol. Assoc.*, **29**, 68–105.

DINES, H. G. and EDMUNDS, F. H. 1933. The Geology of the Country around Reigate and Dorking. *Mem. Geol. Surv.*

DINES H. G, HOLLINGWORTH, S. E., EDWARDS, W., BUCHAN, S. and WELCH, F. B. A. 1940. The Mapping of Head Deposits. *Geol. Mag.*, **77**, 198–226.

——, HOLMES, S. C. A. and ROBBIE, J. A. 1954. Geology of the Country around Chatham. *Mem. Geol. Surv.*

DREW, F. 1861. On the Succession of the Beds in the Hastings Sand in the Northern Portion of the Weald. *Quart. J. Geol. Soc.*, **17**, 271–86.

EVANS, Caleb. 1864. On Fossils from the Railway-cuttings in the vicinity of London. *Proc. Geol. Assoc.*, **1**, 347–51.

—— 1871. On the Strata exposed by the line of Railroad through the Sevenoaks Tunnel. *Proc. Geol. Assoc.*, **2**, 1–4.

FITTON, W. H. 1836. Observations on some of the Strata between the Chalk and the Oxford Oolite, in the South-East of England. *Trans. Geol. Soc.*, 2nd Series, **4**, 103–389.

FOSTER, C. LE NEVE and TOPLEY, W. 1865. On the Superficial Deposits of the Valley of the Medway, with Remarks on the Denudation of the Weald. *Quart. J. Geol. Soc.*, **21**, 443–74.

GALLOIS, R. W. 1966. in *Summ. Progr. geol. Surv.*, *Gt. Brit.*, for 1964, 47.

GOSSLING, F. 1936. Note on a Former High-Level Erosion Surface about Oxted. *Proc. Geol. Assoc.*, **47**, 316–21.

—— 1937. Wealden Pebbles in the Valley of the River Darent. *Geol. Mag.*, **74**, 527–8.

—— 1941a. A Contribution to the Pleistocene History of the Upper Darent Valley. *Proc. Geol. Assoc.*, **51** for 1940, 311–40.

—— 1941b. Field Meeting between Limpsfield and Westerham. *Proc. Geol. Assoc.*, **51** for 1940, 341–5.

GROVES, A. W. 1931. The Unroofing of the Dartmoor Granite and the Distribution of its Detritus in the Sediments of Southern England. *Quart. J. Geol. Soc.*, **87**, 62–96.

HALL, S. 1930. Field Meeting at Tonbridge and Leigh. *Proc. Geol. Assoc.*, **41**, 92–5.

——, MILNER, H. B. and SWEETING, G. S. 1933. A Traverse of the Central Weald. *Proc. Geol. Assoc.*, **44**, 444–7.

HANDCOCK, E. W. 1910. Excursion to Southborough and Tonbridge. *Proc. Geol. Assoc.*, **21**, 521–2.

—— 1914. Report of an Excursion to Tonbridge. *Proc. Geol. Assoc.*, **25**, 56–8.

HARRISON, SIR EDWARD R. 1928. *Harrison of Ightham*. London.

—— 1958. The Riddle of the Old Stones: A personal retrospect. *Archaeologia Cantiana*, **71** for 1957, 47–52.

HERRIES, R. S. 1920. Excursion to Knockholt and Dunton Green. *Proc. Geol. Assoc.*, **31**, 220–1.

HINDE, G. J. 1885. On Beds of Sponge-remains in the Lower and Upper Greensand of the South of England. *Phil. Trans. Roy. Soc.* (2), **176**, for 1885, 403–53.

HINTON, M. A. C. 1910. A Preliminary Account of the British fossil Voles and Lemmings, with some remarks on the Pleistocene Climate and Geography. *Proc. Geol. Assoc.*, **21**, 489–507.

—— 1926. Monograph of the Voles and Lemmings (Microtinae) Living and Extinct. Vol. 1. *Brit. Mus. Nat. Hist.*

HOLLINGWORTH, S. E., TAYLOR, J. H. and KELLAWAY, G. A. 1944. Large-scale Superficial Structures in the Northampton Ironstone Field. *Quart. J. Geol. Soc.*, **100**, 1–44.

HOLMES, S. C. A. 1937. Field Meeting at Wrotham and Borough Green, Kent. *Proc. Geol. Assoc.*, **48**, 350–3.

—— 1962. in *Summ. Progr. Geol. Surv. Gt. Brit.*, for 1961, 33.

HOOD, A. W. 1884. The Chalk of the Medway Valley. *Rochester Naturalist*, **1**, 57–63.

HOPKINS, W. 1845. On the Geological Structure of the Wealden district and the Bas Boulonnais. *Trans. Geol. Soc. London*, (2), **7**, 1.

HOWITT, F. 1964. Stratigraphy and structure of the Purbeck inliers of Sussex (England). *Quart. J. Geol. Soc.*, **120**, 77–113.

HUDSON, R. G. S. and MITCHELL, G. H. 1937. The Carboniferous Geology of the Skipton anticline. *Summ. Prog. Geol. Surv. Gt. Brit.* for 1935, pt. 2, 1–45.

—— and TURNER, J. S. 1933. Early and Mid-Carboniferous earth movements in Great Britain. *Proc. Leeds Phil. and Lit. Soc.*, **2**, 455–66.

HUGHES, N. F. 1958. Palaeontological evidence for the age of the English Wealden. *Geol. Mag.*, **95**, 41–9.

JEFFERIES, R. P. S. 1961. The Palaeoecology of the *Actinocamax plenus* Subzone (Lowest Turonian) in the Anglo-Paris Basin. *Palaeontology*, **4**, 609–47.

—— 1963. The Stratigraphy of the *Actinocamax plenus* Subzone (Turonian) in the Anglo-Paris Basin. *Proc. Geol. Assoc.*, **74**, 1–33.

JUKES-BROWNE, A. J. 1900. The Cretaceous Rocks of Britain. **1**, The Gault and Upper Greensand of England. *Mem. Geol. Surv.*

—— 1903. The Cretaceous Rocks of Britain. **2**, The Lower and Middle Chalk of England. *Mem. Geol. Surv.*

—— 1904. The Cretaceous Rocks of Britain. **3**, The Upper Chalk of England. *Mem. Geol. Surv.*

KENNARD, A. S. 1897. Excursion to Otford and the Holmesdale Valley. *Proc. Geol. Assoc.*, **15**, 209–10.

—— and WOODWARD, B. B. 1897. The Mollusca of the English Cave-deposits. *Proc. Malac. Soc. London.*, **2**, 242–4.

—— —— 1901. The post-Pliocene non-marine Mollusca of the South of England. *Proc. Geol. Assoc.*, **17**, 213–60.

KERMACK, K. A. 1954. A biometrical study of *Micraster coranguinum* and *M.* (*Isomicraster*) *senonensis*. *Phil. Trans. Roy. Soc.*, (B), **237**, 375–428.

KERNEY, M. P. 1965. Weichselian Deposits in the Isle of Thanet, East Kent. *Proc., Geol. Assoc.*, **76**, 269–74.

KHAN, M. H. 1952. Zonal analysis of the Lower Gault of Kent based on foraminifera. *Contr. Cusham Fund*, **3**, (2), 71–80.

KIRKALDY, J. F. 1939. History of the Lower Cretaceous period in England. *Proc. Geol. Assoc.*, **50**, 379–417.

LAMPLUGH, G. W. and KITCHIN, F. L. 1911. On the Mesozoic Rocks in some of the Coal Explorations in Kent. *Mem. Geol. Surv.*

——, —— and PRINGLE, J. 1923. The Concealed Mesozoic Rocks in Kent. *Mem. Geol. Surv.*

LEACH, A. L. 1921. Excursion to the North Downs in Kent. *Proc. Geol. Assoc.*, **32**, 36–42.

LOBLEY, J. LOGAN. 1880. (Report by W. Fawcett). Excursion to Sevenoaks and Tonbridge. *Proc. Geol. Assoc.*, **6**, 202–3.

LOCK, M. 1953. *Equisetites lyelli* (Mantell) at a new Horizon in the Wadhurst Clay, near Pembury, Kent. *Proc. Geol. Assoc.* **64**, 31–2.

MILBOURNE, R. A. 1956. The Gault at Greatness Lane, Sevenoaks, Kent. *Proc. Geol. Assoc.*, **66**, 235–42.

—— 1962. Notes on the Gault near Sevenoaks, Kent. *Proc. Geol. Assoc.*, **72**, 437–43.

—— 1963. The Gault at Ford Place, Wrotham, Kent. *Proc. Geol. Assoc.*, **74**, 55–79.

MILNER, H. B. 1923. The Geology of the Country around East Grinstead. *Proc. Geol. Assoc.*, **34**, 283–300.

MITCHELL, G. H. and STUBBLEFIELD, C. J. 1941. The Carboniferous Limestone of Breedon Cloud, Leicestershire, and the associated Inliers. *Geol. Mag.*, **78**, 201–19.

MURCHISON, SIR RODERICK I. 1851. On the Distribution of the Flint Drift of the South-east of England, on the Flanks of the Weald, and over the Surface of the South and North Downs. *Quart. J. Geol. Soc.*, **7**, 349–98.

NEWTON, E. F. 1937. The Petrography of some English Fuller's Earths and the Rocks Associated with them. *Proc. Geol. Assoc.*, **48**, 175–97.

NEWTON, E. T. 1894. The Vertebrate Fauna collected by Mr. Lewis Abbott from the Fissure near Ightham, Kent. *Quart. J. Geol. Soc.*, **50**, 188–211.

—— 1899. Additional Notes on the Vertebrate Fauna of the Rock-fissure at Ightham (Kent). *Quart. J. Geol. Soc.*, **55**, 419–29.

NICHOLS, D. 1959. Changes in the chalk heart-urchin *Micraster* interpreted in relation to living forms. *Phil. Trans. Roy. Soc.*, **242**, No. 693, 347–437.

OWEN, H. G. 1958. Lower Gault Sections in the Northern Weald and the Zoning of the Lower Gault. *Proc. Geol. Assoc.*, **69**, 148–65.

—— 1963. Some sections in the Lower Gault of the Weald. *Proc. Geol. Assoc.*, **74**, 35–53.

PALFRAMAN, D. F. B. 1966. Variation and Ontogeny of some Oxfordian ammonites: *Taramelliceras richei* (de Loriol) and *Creniceras renggeri* (Oppel), from Woodham, Buckinghamshire. *Palaeontology*, **9**, 290–311.

PRESTWICH, J. 1855. On the Origin of the Sand and Gravel-Pipes in the Chalk of the London Tertiary District. *Quart. J. Geol. Soc.*, **11**, 64–84.

—— 1858. On the Age of some Sands and Iron-Sandstones on the North Downs. *Quart. J. Geol. Soc.*, **14**, 322–35.

—— 1889. On the Occurrence of Palaeolithic Flint Implements in the Neighbourhood of Ightham, Kent, their Distribution and Probable Age. *Quart. J. Geol. Soc.*, **45**, 270–97.

—— 1890. On the Relation of the Westleton Shingle to other Pre-Glacial Drifts in the Thames Basin, and on a southern Drift, with Observations on the Final Elevation and Initial Subaerial Denudation of the Weald: and on the Genesis of the Thames. *Quart. J. Geol. Soc.*, **46**, 155–81.

—— 1891. On the Age, Formation and Successive Drift-Stages of the Valley of the Darent; with Remarks on the Palaeolithic Implements of the District, and on the Origin of its Chalk Escarpment. *Quart. J. Geol. Soc.*, **47**, 126–63.

—— and MORRIS, J. 1846. On the Wealden Strata exposed by the Tunbridge Wells Railway. *Quart. J. Geol. Soc.*, **2**, 397–405.

RAMSAY, A. C., BRISTOW, H. W., BAUERMAN, H. and GEIKIE, A. 1859. *A Descriptive Catalogue of the Rock-specimens in the Museum of Practical Geology.* 2nd Edit. London.

REEVES, J. W. 1968. Subdivisions of the Weald Clay in North Sussex, in Surrey and Kent. *Proc. Geol. Assoc.* **79**, 457–76.

REYNOLDS, S. H. 1909. A Monograph of the British Pleistocene Mammalia, Vol. II, Pt. III, the Canidae. London (*Palaeontogr. Soc.*).

—— 1912. A Monograph of the British Pleistocene Mammalia, Vol. II, Pt. IV, the Mustelidae. London (*Palaeontogr. Soc.*).

SALTER, A. E. 1905. On the Superficial Deposits of Central and Parts of Southern England. *Proc. Geol. Assoc.*, **19**, 1–56.

SEELEY, H. G. 1891. *Handbook of the London Geological Field Class.* London.

SHARPE, D. 1855. Description of the Fossil Remains of Mollusca found in the Chalk of England, Pt. 3. London (*Palaeontogr. Soc.*).

SHEPHARD-THORN, E. R., SMART, J. G. O., BISSON, G. and EDMONDS, E. A. 1966. Geology of the Country around Tenterden. *Mem. Geol. Surv.*

SKEMPTON, A. W. and PETLEY, D. 1967. The shear strength along structural discontinuities in stiff clays. *Proc. Geotech. Cong. (Oslo)*, **2**, 29–46.

SMART, J. G. O., BISSON, G. and WORSSAM, B. C. 1966. Geology of the country around Canterbury and Folkestone. *Mem. Geol. Surv.*

SPATH, L. F. 1923–1943. A Monograph of the Ammonoidea of the Gault. London (*Palaeontogr. Soc.*).

—— and STUBBLEFIELD, C. J. 1930. Note in Circular 331. *Geol. Assoc.*

STRAKER, E. 1931. *Wealden Iron.* London.

TAYLOR, J. H. 1951. Sedimentation problems of the Northampton Sand Ironstone, pp. 74–85; *in* Hallimond and others. The Constitution and Origin of Sedimentary Iron Ores: A Symposium. *Proc. Yorks. Geol. Soc.*, **28**, 61–101.

TERZAGHI, K. 1950. Mechanism of Landslides. *In* Application of geology to engineering practice. *Geol. Soc. Am.* Engineering Geology (Berkey) Volume.

THURRELL, R. G., WORSSAM, B. C. and EDMONDS, E. A. 1968. Geology of the Country around Haslemere. *Mem. Geol. Surv.*

TOPLEY, W. 1875. The Geology of the Weald. *Mem. Geol. Surv.*

—— 1894. *In* discussion on ABBOTT and NEWTON, above. *Quart. J. Geol. Soc.*, **50**, 210–1.

TREACHER, L. 1909. Excursion to Limpsfield and Westerham. *Proc. Geol. Assoc.*, **21**, 59–64.

TRIMMER, J. 1852. On the Origin of the Soils which cover the Chalk of Kent. Pt. 2. *Quart. J. Geol. Soc.*, **8**, 273–7.

WEEKS, A. G. (in press). The stability of natural slopes in South East England as affected by periglacial activity. *Q. Jl. Engng. Geol.*

WELLS, A. K. and GOSSLING, F. 1947. A Study of the Pebble Beds in the Lower Greensand in East Surrey and West Kent. *Proc. Geol. Assoc.*, **58**, 194–222.

WHITAKER, W. 1861. On the 'Chalk-Rock', the Topmost Bed of the Lower Chalk in Berkshire, Oxfordshire, Buckingham, etc. *Quart. J. Geol. Soc.*, **17**, 166–70.

—— 1872. The Geology of the London Basin. *Mem, Geol. Surv.*, **4.**

—— 1908. The Water Supply of Kent. *Mem. Geol. Surv.*

WHITAKER, W. 1912. The Water Supply of Surrey. *Mem. Geol. Surv.*

WHITTARD, W. F. and SMITH, S. 1943. Geology of a Recent Borehole at Filton, Glos. *Proc. Bristol Nat. Soc.*, **9**, 434–50.

WOODS, H. 1896. The Mollusca of the Chalk Rock, Pt. I. *Quart. J. Geol. Soc.*, **52**, 68–98.

—— 1897. The Mollusca of the Chalk Rock, Pt. 2. *Quart. J. Geol. Soc.*, **53**, 377–404.

WOOLDRIDGE, S. W. 1927. The Pliocene History of the London Basin. *Proc. Geol. Assoc.*, **38**, 49–132.

WORSSAM, B. C. 1958. in *Summ. Prog. Geol. Surv. Gt. Br.* for 1957, 29.

—— 1963. Geology of the Country around Maidstone. *Mem. Geol. Surv.*

—— 1964. Written contribution to the discussion of a paper previously taken as read: 6 March 1964. *Proc. Geol. Assoc.*, **75**, 573–5.

—— and THURRELL, R. G. 1967. Field Meeting to an area north of Horsham, Sussex. *Proc. Geol. Assoc.*, **77** for 1966, 263–71.

WRIGHT, C. W. and THOMAS, H. DIGHTON, 1947. Notes on the Geology of the Country around Sevenoaks, Kent. *Proc. Geol. Assoc.*, **57**, 315–21.

M

Appendix

LIST OF GEOLOGICAL SURVEY
PHOTOGRAPHS

(One-inch Sheet 287)

Taken by Mr. J. Rhodes or Mr. J. M. Pulsford

Copies of these photographs are deposited for reference in the Library of the Institute of Geological Sciences, South Kensington, London S.W.7. Black and white prints and lantern slides can be supplied at a fixed tariff, and in addition colour prints and transparencies are available for many of the photographs with numbers higher than 9696.

All numbers belong to Series A.

5351	The Darent Gap looking north from hillside near Broughton House.
5352–3	Views of Chalk escarpment.
5354	Asymmetric fold in Lower Greensand, near Oldbury.
5355	Middle Chalk, Dunton Green Limeworks Pit, Dunton Green.
5356	Gault Clay, Dunton Green Brickworks, Riverhead.
5357–8	Pit in Folkestone Beds, Sevenoaks.
5359–60	Ightham Stone Band (Folkestone Beds), near Oldbury.
5361	Oldbury Rock Shelters (Folkestone Beds), near Oldbury.
5362	Steeply dipping Hythe Beds, near Sundridge.
6807	Fault plane between Wadhurst Clay and Tunbridge Wells Sand, Quarry Hill Brickyard, Tonbridge.
6808	Limestone and sandstone beds in Wadhurst Clay, Castle Hill Brickworks, Tonbridge.
6809	View showing change of slope at junction of Lower Tunbridge Wells Sand and Wadhurst Clay, near Southborough.
6810	Lower Tunbridge Wells Sand overlying Wadhurst Clay, High Brooms Brickyard, near Southborough Station.
6811	Roadside exposure showing cambered Lower Tunbridge Wells Sand, near Speldhurst.
6812	Basal beds of the Ardingly Sandstone near Speldhurst.
6813–4	Landslip at junction of Lower Tunbridge Wells Sand and Wadhurst Clay, near Bidborough.
6815	Jointing in Ardingly Sandstone, near Penshurst Station.
6816	The Chiding Stone (Ardingly Sandstone), Chiddingstone.
6817–9	Exposures of Ardingly Sandstone, near Markbeech.
6821	Hythe Beds in roadside, near Westerham.
6822–3	Quarry in Folkestone Beds, near Westerham.
6824	Phosphatic nodules in Gault Clay exposed in pit, near Westerham.
6825–7	Quarries in Hythe Beds, overlain by Angular Chert Drift, Limpsfield Common.
6828	Quarry in Folkestone Beds, Limpsfield.
6829–30	Quarry in Upper Chalk overlain by Clay-with-flints and sand, near Westerham.

6831–2	Quarry in Folkestone Beds, near Moorhouse Bank.
6833	The same quarry, showing hillwash on Folkestone Beds.
6834	View of Atherfield Clay inliers, near Westerham.
6835	View showing change of vegetation at junction of Hythe Beds and Atherfield Clay, near Westerham.
6836	View showing change of vegetation towards base of Hythe Beds, near Westerham.
7083	Inlier of Ashdown Beds, near Cowden.
7084	Dry Hill Camp, east of Dormans Land.
7085	Landslips at Crockham Hill.
7086	The Weald, looking south-west from the Lower Greensand escarpment.
7087	Lower Greensand escarpment, seen from near Limpsfield church.
7088	Chalk escarpment near Titsey.
7089	Quarry showing fault plane between Wadhurst Clay and Tunbridge Wells Sand, Tonbridge.
7090–2	Landslip at junction of Lower Tunbridge Wells Sand and Wadhurst Clay, near Bidborough.
7093–4	Typical Wealden scenery.
7095	The Weald, from River Hill.
7096–100	Gravel working in River Darent alluvium.
7101–3	Gravel overlying Folkestone Beds in quarry, near Sevenoaks.
7104	Darent Gap looking across Holmsdale from the Lower Greensand escarpment.
7105–9	Quarry in Hythe Beds showing anticlinal folds, Dryhill, Sundridge.
7110	Coombe on Lower Greensand dip slope, Seal.
7111–2	Quarry in Folkestone Beds, Stone Street.
7113	Oldbury Hill, Ightham.
7114	Gorge of River Bourne, Ightham.
7115–6	Quarry in Hythe Beds with capping of Sandgate Beds, Ightham.
7717–9	Quarry in Folkestone Beds, Ightham.
7120–7	Quarry in Folkestone Beds, overlain by Gault and gravelly brickearth, Borough Green.
7128	Folkestone Beds, Rock Tavern, Borough Green.
7129	Crow Hill (Folkestone Beds), Borough Green.
7130	Folkestone Beds in railway cutting, near Borough Green.
7131	Shallow working in Gault Clay, near Borough Green.
7132	Quarry in Folkestone Beds with well-developed ironstone, Borough Green.
7133	Quarry in Folkestone Beds, Platt.
7134–8	Quarries in Folkestone Beds, Borough Green.
7139–40	View across Gault Clay vale to Chalk escarpment, from near Borough Green.
7141–2	Quarries in Middle Chalk, near Wrotham.
7143–5	Quarries in Folkestone Beds, Wrotham Heath.
7146–8, 7156	Views of Chalk escarpment.
7149–50	Views across Holmsdale, from the Chalk escarpment.
7151–2	Panoramic view of Chalk escarpment and Gault Clay vale.
7153–5	Quarry in Folkestone Beds, Ightham.
7157–8	Quarry in Sandgate and Hythe Beds, West Malling.
7159–60	Valley cutting into Hythe Beds, West Malling.
7161	Coombe in Hythe Beds, Offham.
7162–5	Quarry in Hythe Beds, Offham.
7166	Anticlinal fold in Hythe Beds, Offham.
7167–70	Quarries in Folkestone Beds, Trottiscliffe.
7171–2	Quarry in Folkestone Beds, Ryarsh.
7173	Quarry in flinty drift, Ryarsh.

7174	Alluvial flat and steep valley side, Ryarsh.
7175	Quarry in Gault Clay, Kemsing.
9146–7	Faulted junction of Tunbridge Wells Sand and Wadhurst Clay, Quarry Hill Brickworks, Tonbridge.
9281	Quarry in gravelly drift, with disturbed Gault and Folkestone Beds, Riverhead.
9697	Escarpment of Hythe Beds, Ide Hill.
9917	Wadhurst Clay, Quarry Hill, Tonbridge.
9919–20	Sand dredging of Folkestone Beds, Riverhead.
10269–70	Anticlinal structure on Hythe Beds, Dryhill Quarry.
10271	View towards Penshurst from below the Lower Greensand escarpment.
10273	Ardingly Sandstone near Markbeech.
10274	Quarry in Ardingly Sandstone near Penshurst Station.
10275–6	The Chiding Stone, Chiddingstone.
10277–9	Quarry in Hythe Beds, Offham.
10280	Solution pipes of Head in Middle Chalk, near Brasted.
10281	Quarry in Middle Chalk, Dunton Green.
10282–5	Pit in Folkestone Beds and Gault Clay, Westerham.
10309–10	Quarry in Lower Tunbridge Wells Sand and Wadhurst Clay, Southborough.
10377	Palaeolithic Rock Shelters; sandrock of Folkestone Beds, near Ightham.

INDEX

N

Dd. 142383 K12